THE PROBLEM OF HOMOSEXUAL BEHAVIOR

and

Cyclopedia of Normal and Abnormal Sexual Behavior

THE PROBLEM OF HOMOSEXUAL BEHAVIOR

and

Cyclopedia of Normal and Abnormal Sexual Behavior

by

The Society for the Scientific Study of Male Psychology and Physiology

Edited by Jerry Bergman

Creation Summit Publishers
LINDENHURST, ILLINOIS

Published by
Creation Summit Publishers
Lindenhurst, IL 60046
https://creationsummit.com/

Published 2026.

Printed in The United States.

Distributed by
Amazon.com

Cover and book design by Richard Geer.

The Society for the Scientific Study of Male Psychology and Physiology
The problem of homosexual behavior and cyclopedia of normal and abnormal sexual behavior. Edited by Jerry Bergman

Includes bibliographical references.
ISBN: 979-8-9950778-1-7 (softcover).

Contents

Acknowledgements

I want to thank: David Demick, M.D.; Bert Thompson, Ph.D.; Susan Loeffel, M.D.; Wayne Frair, Ph.D.; Donald Tyler, M.D.; Tom Baker, M.D.; John Woodmorappe, M.A.; Dianne Bergman, and Ernie Johanson, Ph.D., for their comments on an earlier draft of this book. All of the conclusions are the author's and not necessarily those of the reviewers of the publisher.

PART I:
THE PROBLEM OF HOMOSEXUAL BEHAVIOR

Introduction to Part I

The propriety of homosexual behavior has been a topic of much discussion for several decades. At the bottom of the controversy in the religious community is the scriptural condemnation of homosexual behavior. This monograph argues that a major reason why the Scriptures condemn homosexual behavior is its long-term detrimental effects on health. A review of the medical literature focusing on males documents that homosexual behavior has clear detrimental effects on health, causing scores of serious (and lethal) diseases. As a result, one study found that persons of either sex involved in a lifelong homosexual lifestyle live, on average, only into their 50s. Although recent effective treatment of AIDS (Acquired Immunodeficiency Syndrome) has mitigated some of the data reported here, most of the other conclusions are still valid.

Note: This review serves only as an introduction to the health concerns of homosexual behavior and related. As such, consultation with your doctor or other health professional is required for the latest information, diagnoses, and treatment of all medical conditions.

CHAPTER 1

Defining 'Homosexual'

From the medical standpoint, 'homosexual' is not a person, but rather a behavior. Many self-labeled heterosexuals, at times, involve themselves in homosexual behavior. Conversely, many persons self-labeled as homosexual involve themselves in heterosexual behavior. Despite decades of research looking for a homosexual gene, no gene has been identified that explains homosexual behavior, nor one for heterosexual or even bisexual behavior. As is true of all sexual behavior, both positive and negative early experiences are critical. This is especially true of homosexual behavior. The term 'gay' used to mean happy; now it means homosexual and implies that homosexuals are happy, and heterosexuals are not. This is demeaning to those who have the straight label! The term is often used now because it is less formal and more friendly.

Case example: The problem with the homophobia claim is: "I was homosexually molested, actually raped, when I was twelve, twice by the same man, a close friend of the family. My psychiatrist put me in a support group for those who had the same experience I had. So, my perspective is beyond my own. This problem is really common. In my case, I have a real fear of homosexuals, and this fear has transferred to all men. I am very careful about who my male associates are and try to avoid men who could put me in a compromising position. If a woman were raped several times and, as a result, became fearful of men, dating men, and so on, would you call her a heterophobe? To me, homophobia is a normal reaction to being molested, and when the term is used as it is today, it impugns the victim."

For this reason, the term homophobia is inaccurate and inappropriate. A more accurate term, such as homosexual prejudice, is much better. Also, millions of boys have been homosexually molested, so this is by no means a small problem.

THE NUMBER OF PERSONS SELF-LABELED AS HOMOSEXUALS

Numerous studies have been completed to determine the percentage of the American population that claims to be homosexual. The best research has determined that it is about two to three percent of all males and one to two percent of females. Due to the influence of the current gay rights movement, the number may be higher now.[1] Research has found that very close to this percent also reside in Europe, Canada, and Australia. Because humans tend to judge by our limited experience, large systematic scientific surveys are required, as was done in one recent study, which debunked the 'born that way' myth.[2] The fact is, "Some guys in the military and in prison get involved in homosexual sex and, when they get out, often go back to women. If you label them gay, they will knock your block off."

Problems in obtaining a more accurate number include many people are bisexual or non-binary but answer "gay" in surveys, which makes it difficult to obtain accurate data. The commonly cited number, ten percent, is from the Kinsey research completed in the late 1940s, which has now been refuted. Alfred Kinsey was the famous sex researcher who wrote the bestselling book *Sexual Behavior in the Human Male* and, a few years later, published *The Sexual Behavior in the Human Female*. They were published by Sanders, a prestigious medical publishing company, and each was about 500 pages in length.

Another problem is it is patently false that homosexuality is a uniform attribute across individuals. It is also false that it is stable over time and that it can be easily measured—a large systematic scientific survey sampled over 100,000 individuals across the globe. One controversial conclusion was that homosexuality, as defined by scientifically rigorous criteria, *tends* to 'mutate' into heterosexuality over the course of a lifetime, especially in women. As one man commented, "I like children, and holding one makes me think, as a gay man, would I ever have a child? It's something we think about at times. Life has been very fulfilling so far, but I came from a very family-oriented environment and have to think, could my life be better?"

CAUSES OF NORMAL AND ABNORMAL SEXUAL BEHAVIOR

The cause of both normal and abnormal sexual behavior is, in most cases, both positive and negative early sexual experiences. This is why most people, as far as possible, attempt to shield children from sexual behavior, a problem in the internet

---

1 Diamond, Lisa. 2015. "Time to lay 'born this way' to rest." *New Scientist* **227**(3031):18-19, July 25.

2 Diamond, Lisa. 2015. "Time to lay 'born this way' to rest." *New Scientist* **227**(3031):18-19, July 25.

age that broadcasts sexuality everywhere, both normal and abnormal. The latest study on the influence of genes as a causative factor in homosexuality has confirmed what we already knew from numerous previous studies. This, the largest-ever study of the role of genes in homosexual behavior, found no evidence of a genetic cause of homosexuality. The study included more than 470,000 people. The conclusions include five genetic variants that are associated with having a same-sex sexual partner. Those variants did not predict people's sexual behavior, but those five variants point to biological processes involved in choosing sex partners.

For instance, one variant identified in the study has been linked to male-pattern baldness, and another to the ability to smell certain chemicals, which may affect sexual attraction. The five statistically significant variants explain less than one percent of homosexual behavior. The major environmental influences include an array of social and cultural factors that affect behavior. In short, the study's lead author, Andrea Ganna, a geneticist at MIT and Harvard University, published in *Science* concluded, "There is no 'gay gene' that determines whether someone has same-sex partners."

Simon LeVay and Dean Hamer's research[3] has been widely criticized.[4] This and other studies have not been replicated. In medicine, until a study is replicated several times, the results are usually considered tentative at best. Research has found that humans **can** inherit traits such as certain personality characteristics, which in rare cases may predispose one to homosexuality.

Some research indicates that the cocktail of hormones in the womb and other factors in the womb environment, plus factors such as stress during pregnancy and environmental factors after birth, all may contribute to sexual orientation. This causation problem is very complex, and this is why many different opinions exist about the cause of homosexuality. We know that the gay lifestyle is a major reason for their health problems, but we also need to look at factors that affect the embryo and fetus before birth. The fact is, gays drink more, smoke more, and in general take more risks in life. We should not excuse bad behavior, blaming it on the

---

3 LeVay, Simon, and Dean H. Hamer. 1994. "Evidence for a Biological Influence in Male Homosexuality.
Two pieces of evidence, a structure within the human brain and a genetic link, point to a biological component for male homosexuality." *Scientific American* 270(5), May. https://www.scientificamerican.com/article/evidence-for-a-biological-influence/.

4 Byne, William. 1994. "The biological evidence challenged." Scientific American 270(5):50-54, May; Horgan, John. 1995. "Gay genes, revisited: Doubts arise over research on the biology of homosexuality." Scientific American 273(5):26, November; Santinover, Jeffrey. 1996. Homosexuality and the Politics of Truth. Grand Rapids, MI: Baker Publishing.

mother as some people try to, but these factors need to be evaluated. An example is height, which is inherited helps greatly to be a successful basketball player. Because a trait has a genetic component does not mean it is genetically caused. Basketball requires skills that have a genetic component. A common one is tallness, but we would not call the ability to play basketball a genetic trait. All human traits have one or more genetic components. Think of football (body size is important), or track (running speed is central to success which depends on certain body traits) but none of these sports is genetically caused.[5]

If there exists a gay gene, there must be a gene for bisexuality. Is there is a gene for heterosexuality, and another one for bisexuality? What about pedophilia or necrophilia? We have not found a gene for any sexual preference. Not for heterosexuality, bisexuality, or homosexuality, or for any other philia. Genes exist that produce males and females, and various sexual traits, such as hair color, but no documented gene is known for any sexual orientation. Our best guess is that genes, hormones (especially during early development in the womb), and also the environment once out of the womb, may all be important, but genes are usually the least important cause compared to these other causative factors, especially early life experiences.

Sigmund Freud concluded that everyone goes through a homosexual developmental stage as an adolescent. He concluded that people who are gay as adults never progressed beyond the homosexual stage of development. They are labeled fixated or stuck at some early developmental stage. One clinical professional concluded, "I never found any reason to question that conclusion. Most people do not act out sexually during this homosexual stage, but feelings for the same sex are there, even if they are usually repressed." Freud made no claims that this stage was based on genetics but was part of the life stages we all progress through.

IS THE FAMILY OF ORIGIN IMPORTANT?

One study done in Canada, using a 20-percent sample of the entire Canadian census on children reared in same-sex households (gay households), two men as parents or two women, found that their children were only 65 percent as likely to graduate from high school compared to the traditional family, and that daughters did worse than sons. They did worse in other ways as well. The study did not research why the difference existed, which will be the topic of other research.

5 Santinover, 1996.

ASEXUAL PERSONS

Asexual persons are not attracted to either males or females. They have no concerns about sexually transmitted diseases. Asexual people make up an estimated five percent of the population but are totally ignored by almost everybody. For an asexual, sex with anyone is not so much repulsive as it is baffling. Most never dated, never wanted to, and never married, so some people assume they are either homosexual or heterosexual. Asexuals don't go marching in the streets or get the newspaper headlines as some people do. Most are very happy as they are, and work hard and often have done well in life. They don't have to worry about divorce, dating, or sex. Some believe that the problem is that they have a very low sex drive, so low that it's nonexistent. Asexuals also do not relate to people sexually. They see a man or woman more as another person than a male or female gender. People ask them, "Why aren't you married? Are you gay?" Asexual's learn it is often difficult to explain their sexual orientation, so they often don't try. He may explain, "I have no desire to get married, or even have a girlfriend," but many asexual's marry for companionship.

INTERVIEW OF A BISEXUAL MALE

The client was married for over a decade and had three kids. One of our other employees, an openly bisexual woman, invited Susan to celebrate her birthday. They got together several times after that, and Susan soon developed strong feelings for her. Susan asked the woman out several times and was told she was busy. A few days later, Susan saw the woman in a restaurant with another woman, holding hands and even kissing. She became very jealous and wanted to tell her that she had very strong feelings for her but decided not to. I wondered how a straight woman could become bisexual so easily. This goes against my belief that people are born straight, bisexual, or gay. Maybe Freud was right. We all have latent homosexual feelings which must be developed, or they are extinguished if we grow out of them!

Another client was a divorced woman who explained, "The dating section of the want ads noted that most of the ads were like the following one, which reads 'Make a Bi-Curious Connection Call' and then lists the phone number. The next one read, 'Meet Gay & Bi Locals. Browse & Respond for Free,' then gave the phone number. Why do they label themselves and then use the label to condone their behavior choices? If you ask me, these are just a bunch of men looking for kinky sex, like my ex-husband."

PROMISCUITY

Some gay couples live together into their old age, but this is unusual according to the gay sex columnist Dan Savage, who wrote, "Some gay couples believe that if you're in love, you're not going to want to have sex with anyone else. That's not true. We need to acknowledge that people don't have to spend forty years of marriage lying to and policing each other. So, the open marriage option is not only tolerated, but is seen by us as realistic, and is actually encouraged in the gay community."

This conforms to the scientific literature. Blumstein and Schwartz interviewed hundreds of American couples and found that gay men have sex more often in the early part of their relationship than do heterosexuals, but after about a decade, they have sex less frequently than heterosexual couples. The researchers found that the main problem was that interest in their partner declined, while interest in homosexual sex remained high – thus sex with other men became increasingly important as time went on. This trend was also found in stable homosexual relationships. They also found that most gay men do not care if their partners are not monogamous. A gay man who is monogamous is such a rare phenomenon that he may have difficulty being believed.

Blumstein and Schwartz also found that there was a big difference between gay males and lesbians – males seek much more variety and seldom stop seeking outside partners. Conversely, lesbians have fewer outside partners than do all the other groups the researchers studied. Lesbians are more interested in sex with a person who loves them, but gay men are more interested in sex with someone sexually desirable. Being in love with them was not nearly as important.

They also concluded that lesbians are not comfortable with the role of sexual aggressor, and this was one reason why they have sex less often than heterosexual couples. Conversely, for gay male couples, both males tend to feel free to be the initiator, which, Blumstein and Schwartz concluded, can create problems in gay relationships. As one of our clients stated, "Many gay men feel they are gay and proud, and to live like a straight man is like being a traitor to the cause. *Being gay is like being loyal to your nation, like being an American.* We are a community and have to stick together. We have to survive!"

Although many exceptions exist, the proportion of people who adopt a homosexual identity and the length of time they maintain it is primarily a result of environmental factors identifiable in these epidemiological studies. These factors – both 'cultural' and 'demographic' – include things such as early sexual experiences, childhood sexual abuse, social networks, education, and cultural beliefs. The study is not about men with certain hormone issues or pseudo-hermaphrodites. As to the claim that homosexuality mutates into heterosexuality, some claim that the opposite may actually be more common.

CHAPTER 2

The Problem of Molesting Children

A 2004 research study by the John Jay College of Criminal Justice for the United States Conference of Catholic Bishops identified 4,392 Catholic priests and deacons who were in active ministry between 1950 and 2002 and who have been accused of underage sexual abuse of 10,667 individuals. The fact is, the Catholic church for decades openly accepted homosexuals into the ministry because all priests, both gay and straight, took vows to be celibate. Thus, their sexuality was not a concern until the lawsuits began increasing. The cost to the church so far has been over four billion dollars. The Boy Scouts have also faced the same problem. A total of 7,819 male volunteers were removed as a result of evidence that they sexually abused male children, forcing the Boy Scouts into bankruptcy due to the lawsuits they faced as a result of the abuse.

GAY PRIESTS

Case history of a Catholic priest who is gay. "Many gay men become priests. We all must take a vow of chastity and, therefore, whether we are gay or straight does not matter. We are all chaste. The church has no problem with our gay past because we are now all chaste." Some gay priests break their vows, as do some straight priests, which we hear a lot about. In seminary, I was by no means alone as a gay man, and I can tell you that almost all of them are celibate. I lived with these guys for four years in seminary and got to know them very well. Some have struggled with their sexuality and left the priesthood, but most were determined to follow their vows. We were very supportive of each other and became like brothers. And most of us were very dedicated to each other and to our church."

"I became a priest for the same reason most of my fellow priests did: our love of God and the Church and our desire to serve others. As a gay man, I saw that many of my fellow gays lived somewhat irresponsible lives. I am not sure why, but before I decided to become a priest, I earned a master's degree in social work and was very interested in the gay issue. The scientific literature is very clear. We often live very reckless lives. We smoke and drink more than the straight population, and I saw this same behavior in my gay friends. We are also often very promiscuous and engaged in risky behavior, such as using drugs to increase our libido. I was no exception. One day, when my third closest friend died, I looked at myself and thought, 'I have to change my life. I cannot live this life anymore.' I eventually became a priest and have never regretted it."

"Many of my friends were also not reckless. You cannot condemn the many for what the few do. But far too many gays are reckless as I once was, so for me the priesthood was the perfect answer. And I may add that many of my brother priests mirrored my experience. I do not feel a need to indulge in my past life and am very happy with my life now. I just wanted to tell my story. This all came back when my best friend from childhood died from some mysterious disease that caused his T-cells to slowly disappear, from the disease you referred to, maybe AIDS."

BEING REARED IN A GAY FAMILY

A Swedish study on children who grow up in gay families found that children whose parents divorce were more likely to marry homosexually than those growing up in intact families. Childhood correlates of first marriages were researched in a national cohort of two million, city-reared, 18-to-49-year-old male Swedes, a country where homosexuals can marry. We found these children were significantly less likely to marry heterosexually but more likely to marry homosexually than their rural-born peers. Heterosexual marriage was significantly linked to having young parents, small age differences between parents, stable parental relationships, and late birth order.

For males, homosexual marriage was associated with having older mothers, divorced parents, absent fathers, and being the youngest child. For women, maternal death during adolescence and being the only, or the youngest, child, or the only girl in the family, increased the likelihood of homosexual marriage. This study provides population-based evidence that childhood family experiences are critically important determinants of heterosexual and homosexual marriage decisions in adulthood. This study clearly shows that homosexually-oriented children tend to be innocent victims of bad and unstable families, with other variables, such as older brothers and older mothers, thrown in.

While 'free will' is involved in every human act, and therefore there is, to some degree, a moral dimension to homosexuality, the real issue is about long-term happiness and well-being. Children, like those in the Canadian study, don't 'choose' their gender identity and sexual prospects, like the kids who were raised in what are apparently more stable families.

PEDOPHILIA

Anthropologists who have studied the Afghan culture for over thirty years, have evaluated in detail the culture in which older men took 9-to-15-year-old boys as lovers. This practice is very common in this culture. One estimate is that about half of the Pashtun tribal members in southern Afghanistan take young boys as lovers, a practice called bacha baz, meaning boy player. This gives the older men status, and they can then enjoy boasting about their bacha baz to their friends. Young boys often dress up as girls, using make-up and bells at dance parties. The men yell things and leer at the boys, throw money at them, and sometimes take them home for sex. In our Western culture, we view having sex with 'dancing boys' as pedophilia rape, a serious crime.

Some Muslim cultures condemn homosexuality and even stone homosexuals to death, but do not regard this practice as homosexuality. It is merely sex for pleasure. They believe that homosexuality involves love, but bacha baz involves only sex, not love. Even after marriage, the men often continue their bacha baz. They believe women are for having babies, and young boys are for pleasure.

Their Imams focus on the passages on menstruation in the Koran that teach that women are unclean. The men argue, "How can one feel desire for a woman whom God made unclean? Besides, women are hidden away, covered from head to toe with a burka. Women are allowed to go out of the house only with their father, brother, or their husband, so men rarely see much of women outside of their close family. As one man proffered, 'we can see the boys' faces but not the women's faces, so we can tell which boys are beautiful.'"

CHAPTER 3

Do More Than Two Different Genders Exist?

The many anatomical, physiological, and psychological differences between the two genders, male and female, are a major part of all college anatomy courses. An individual woman may be taller than some individual males, but the average male is taller than the average female. So, it is with every other known human trait. In fact, for every trait so far examined, the average male is anatomically and physiologically different from the average female. On average, women have a wider field of peripheral vision than males, meaning they can see more around the sides of their direct vision than men. Men, on the other hand, have higher visual acuity and spatial resolution, meaning, on average, they can more effectively track moving objects and see fine details at a distance than females.

I once had a client at the psychological clinic where I was working then who loved masculine outdoor activities, such as chopping wood and mowing the lawn. She had no interest in most activities that females enjoyed, such as cooking, knitting, sewing, needlework, and quilting. Consequently, she did not fit in with her female peers, who thought she was strange. We helped her realize that part of her problem was that she needed to find some new friends. She attended various single get-togethers at local churches and met a male who enjoyed many of the activities that she did. She is now married and doing well, enjoying many outdoor activities with her husband.

Although the above resolution may be obvious to most readers, it is not obvious to a teenage girl attending a small country school facing open rejection from her peers. It is also not obvious to many in the transgender movement who ignore the many well-documented differences between males and females. If a male has a few traits that are close to a female, he is not a female because the vast majority of his traits are male traits. Society has accepted this fact since Adam and Eve.

Thus, we have girls who are tomboys and boys who enjoy typical female pursuits, such as crocheting, knitting, and quilting. Girls who are tomboys are not boys. Some people in the transgender movement appear to be aggressively pushing the idea that only two genders exist is a myth. An example is Theis, a licensed clinical social worker who filed a lawsuit against an Oregon school district recently after officials banned him from displaying behind his desk two children's books titled He is He and She is She.

The books are about God's design for males and females. They explain to children ages 2-8 how they can embrace the way God made them. Theis exclaimed that the books offer a positive and encouraging message to children, teaching that it's good to be a boy, and it's also good to be a girl. He noted that kids need this affirmation at a time when there are a lot of other confusing messages existing in society.

Three weeks after he displayed the books, the middle school's principal informed Theis that an employee complained that the books were "transphobic." The principal noted he did not find anything offensive or inappropriate in the materials, but the employee filed a bias incident complaint against Theis, requiring the district to investigate to determine if the books were "a potential bias incident relating to another person's gender identity."

Officials ultimately determined the books were "a hostile expression of animus toward another person relating to their actual or perceived gender identity." The problem was that the books "promote a binary view of gender, which excludes and invalidates an understanding of gender diversity." They ordered Theis to remove the books or face termination.

Officials justified their actions by a school district policy that allows them to censor expressions they deem target a certain group, in this case, transsexuals. Theis's appeal was denied, so he filed a lawsuit to support his free speech rights.

His attorney believes the school speech policy unconstitutionally limits speech because the district allows other employees to decorate their spaces with other political and cultural messages. Theis has a First Amendment right to display books that exhibit his views. The school library has books on display that include opposing viewpoints, including on homosexuality, a back-cover image of two boys kissing.

His attorney added that the school encourages teachers to express messages about gender that the school agrees with, but censors opposing views. The Supreme Court ruled that the government can't punish employees for expressions that officials disagree with. The fact is, many schools allow staff members to express only one view on gender. An Indiana high school music teacher was forced to resign due to his religious beliefs on gender. Specifically, he requested a religious accommodation to refer to students by only their last names to avoid referring to them by names or pronouns that didn't match their biological sex.

Two female educators in Oregon argued before the 9th U.S. Circuit Court of Appeals for their right to express views on gender in an organization that operated outside school hours. School officials fired the educators after several staff members labeled them "anti-trans." The problem is that students today are being told that their environment or their feelings determine who they are rather than their biology. This issue is most acute in sports, specifically involving biological males competing in women's sports in violation of the President's order that only biological women are allowed to compete in women's sports. This is another example of another problem that has now ensued from ignoring settled science and the core teachings of Judaism, Christianity, and Islam.

CHAPTER 4

Case Histories: Becoming Gay

"Is sexual orientation fluid?" "Yes, there is no question, it is. Let me give you an example that supports the fluid theory, which is typical of those I have included in my new book on sexual fluidity. Rachel, now employed as an engineer by a large firm in Maine, dated primarily one man in high school. Although he wanted to marry her, she evidently never had enough respect for him to marry him, so they broke up. She later met a young man at work whom she was immediately attracted to. They dated and, in a time span that many of us regard as far too short, became engaged. At this time, he was called for active duty (he was in the Army Reserves) and was sent to Afghanistan. Romantic letters flowed freely between them for the next year until, all of a sudden, he stopped writing to her. Concerned that he had been killed or injured, she inquired of the military – and found out that he was still alive and active in the Army."

"Finally, she received a letter that basically stated that he wanted to break off the engagement because he had found someone else. She never found out who this someone else was. Obviously very upset, she went through a time of great bitterness. In the decade that followed, she has not married or dated and has recently declared that she is a lesbian. She has been romantically involved with at least one woman who is also considerably overweight, as she is now. She was well-liked among her coworkers, but some felt that she acted similar to the scorned woman in Charles Dickens' famous novel *Great Expectations*, although she claimed that she is happy with her life now. I have collected scores of similar cases that document my thesis – the fact is, sexuality is fluid. We are all born with certain potentials, but they must be shaped by the environment to mature into gay or straight."

Case history: "I think we are all bisexual. In college, I ran around with a group of wild guys and gals, and it seemed that almost everyone tried bisexuality. The gals really got into it. Most gals can go either way, at least judging from my experience and what I have read. I had a girlfriend that kept a tight rein on me, so I didn't get too involved in the bi world, but many of my friends did. I frequently got together with the other women from work to socialize. There were about thirty-five of us, and we all worked in the entertainment business, so we were all very liberal about sex. We agreed that many women can go either way, and some of us did. So, I think most people are bi, at least most women are if they give themselves a chance."

"A professional photographer began doing weddings and portraits. It was hard to support a family on that work, so I moved on. I freelance now and do a lot of pornography. I have three kids that I want to send to college. Anyway, girl-on-girl sex is big now, and most of the stars do it without much persuasion. I asked them if they enjoyed it, and almost all strongly claimed they did. It was different, and they implied they liked some variety. If you ask me, most women are bisexual. I have talked to many other photographers, and they all agree with my conclusions."

"The main thing the girls have in common is they are all beautiful and willing to be exhibitionists to make some money and have a shot to become famous. They come from all backgrounds – rich, poor, educated, uneducated, smart, and not so smart. Like anything else, they get used to it. Many are shy at first, but most like the attention, and they get it by displaying their body for the world to see. The money is good, and many see it as a route to movie stardom, although very, very few become stars. They tell me, at the least, the work is better than working as a waitress. Another opinion is that people who claim to be bi are just confused. Maybe they have not grown up and are still experiencing their teenage hormone surge."

Case history of a man adopting the gay lifestyle told by a heterosexual female friend: "I am from a small town, and since junior high, I had a crush on Tye, a shy, retiring young man. Later on, he became my steady boyfriend. He was not shy about sex, I could add. When we graduated from high school, we each went our own separate ways. I tried to keep in touch with him, even after I married. He decided to major in international relations in college and went to Australia to study. Once there, he had a difficult time adjusting to the big city. Although shy, he was a very social person and felt extremely lonely in Australia. Finally, he stumbled upon some guys at school who befriended him. He later found out that his new friends were part of a gay group, but he liked them and recognized his strong social needs at this point in his life. Once, he went with them to a gay club."

"It turned out that he not only liked but enjoyed the company of the men at the club enormously. Although never abnormally aggressive sexually, he became very comfortable with and slowly became part of the Australian gay scene. When he came back to the United States, he continued to live this lifestyle and, in time, he 'came out,' declaring that he, too, was gay. I asked him, his partner, and their best friends and their partners if they thought they were born gay. About half claimed they were, the other half insisted they were not. Interestingly, the partner pairs were evenly divided on this question! My former heterosexual boyfriend claimed he was not born gay."

One contributor to this book observed: "In the anthropology class that I teach, some type of fieldwork is required. Unfortunately, most of the students are not able to fly down to Brazil to spend time studying the Amazon Plains Indians, so we are forced to study local subculture groups. Favorite topics in the past have been Hispanics, African-Americans, the Ku Klux Klan, and various religious groups. Invariably, one or two students do their fieldwork on the gay community, typically because they grew up with someone who later adopted the gay lifestyle."

"The students endeavor to determine what factors motivated the adoption of the gay lifestyle, focusing especially on the gay culture. It is a complex culture, with strong support groups. When stranded in a city while traveling, they will often help each other. Just call the local gay bar and you have a whole world of instant friends."

"One student wrote that her best friend, whom I have known for my entire life, was constantly picked on by her older brother and her friends. They accused her of being too tomboyish and, in fits of anger, sometimes called her a fag or some other derogatory name. She dated very little in high school and never had a steady boyfriend. A year after she graduated from high school, she said to me, "I think I am a lesbian." I have known her since we were both very young, and I have concluded that the teasing and labeling influenced her to adopt a lesbian self-image and, eventually, this lifestyle."

"My friend was very tomboyish. Being a tomboy was not unusual for a girl then, so I wondered why she adopted the lesbian lifestyle when most of her friends did not. She also never adopted a feminine persona. I think part of the reason was that she had several really bad experiences with guys, and her brother was downright sadistic to her. She eventually developed a love-hate relationship toward guys. She talked about boyfriends and sex like all of my friends did in high school. Only after she was a year out of high school did she express the idea that she was a lesbian. It was a couple more years before she announced to the world that she was a lesbian and 'came out,' as they say. She found enormous acceptance in the lesbian community – even the gay guys treated her well."

"I, too, was a tomboy, and people tell me I was very difficult to get along with. Nonetheless, people consistently gave me feedback that I am a nice person, even attractive, and I have never had a problem getting dates. People used to compare me to Sophia Loren, although I am not quite as heavy as she was. I am Italian, so there is some similarity. In my case, my two older brothers were very protective of me, their 'little sister.' I was the only girl in the family besides my mom and was treated like a queen at home and elsewhere."

"From this comparison, it is not hard to understand why one girl adopted the heterosexual lifestyle and the other girl adopted the homosexual lifestyle. We are only in our early thirties now, but both of us are very set in our ways. One has now become more masculine – she tries to dress in a masculine way, very short hair, no makeup, rarely wears a dress, and she is now a few pounds overweight, although not unattractive."

The case history of William: "Let me tell you my story. I grew up knowing my uncle fairly well and became aware that he was a homosexual when I was very young. He was a well-educated engineer, had a good job, and lived with his wife in a large, beautiful house. We used to visit them fairly often. He was a very likable person, and, as I remember, a very good-looking man, although somewhat nerdish with his large glasses and an inability to explain himself very well. He married later in life – at that time, I was fourteen, and I judged his age as probably in his late thirties. They did not have any children but adopted a little boy."

"When I asked my folks about their little boy, they implied that, since my uncle was a homosexual, he did not have sex with his wife and, therefore, they had to adopt a child. Curious about such matters, I asked around about their situation. As far as the family knew, no one knew for sure if this was why they adopted a child – we just assumed it was the reason. They could have adopted a child for many reasons, possibly purely altruistic or because of fertility problems."

"When their child was about five, my uncle died of a heart attack while driving into his garage, wrecking his car and the front of the garage. His wife came out of the house screaming and called the EMS as soon as she realized what had happened. It was too late. I remember we visited them shortly after the accident, and she told us the story of how he died. About two years later, she committed suicide."

"Her relatives explained that she found it increasingly difficult to cope without her husband. They were very compatible and very much in love. Although we all knew she was very much in love with him, we assumed that their child would give her a strong reason to live. Obviously, he did not. We lost track of him, but I was often curious to learn how their child had been doing. He must now be in his thirties."

"This example occurred when there was more pressure for heterosexual marriage. Nonetheless, it was apparent to all that they loved each other very much and got along extremely well. We felt it was inappropriate to inquire about their sex lives, but even if they had no sex life, it was clear that they were better off than most couples, including many of those who had an active sex life together. The family was rather open about his homosexuality, and this was never questioned. Evidently, there were some arrests in his past connected with his behavior. His wife also knew about his past and was willing to deal with it. It would have been insightful to interview them in depth to learn more about their relationship. If he were alive today, he would likely have 'come out' and lived a gay lifestyle, but I wonder if he would have experienced the satisfaction in life that he did."

Case history: "One man came out of the closet and left his wife and two young boys. They were divorced, and he moved in with his lover, one of them, that is. He is now back with his wife and boys. He told me that he realized that what he did hurt his boys enormously, and he now realized that they were more important than his sex life. It was difficult for his wife, but they always got along very well, so she agreed to let him move back in. He felt that this decision was best for everybody involved."

Gloria case history: "I have a friend from church that I have been seeing for two years. I am a young widow, and my friend never married. We are both retired and not in need of money. My friend told me he is gay, so romance is out of the question. We do a lot of things as a couple and have a wonderful time together. We eat out often, go to shows, go shopping together, and are in each other's company almost every weekend and often during the week. I made it clear that I wanted to be good friends with him and that's all, and he feels the same way."

"The problem is his father. I love the man, his father, that is, and am closer to him than I was to my own dad, who passed away a few years ago. Ironically, the problem is that I went to a nightclub with some girlfriends from church to hear my favorite musical group. My boyfriend knew I was going with my girlfriends, did not drink, and that we girls left together by ourselves. My boyfriend's dad found out about my visit to the nightclub, and since then has forbidden his son from seeing me. I am no longer welcome in his home, but my friend sneaks out – he lives with his dad and is his caregiver – so we still spend a lot of time with each other, but the lying required for us to see each other bothers me a lot."

The therapist's response: "Your friend needs to make it clear that you did nothing wrong and that, after all, Christ ate with sinners. I feel going to a nightclub with girlfriends from your church to listen to music is not a major problem. I guess that he might be jealous of you, as he now has some competition for his son's attention. I am sure he will come around in time."

CASE HISTORIES

Two basic contrasting theories exist on what causes homosexuality. One is that the cause is due to heredity, the other is that homosexuality is caused by the environment.

"One personal account was a person who has six friends who are gay, and, at a get-together last week, I asked why. Three replied they were born gay, and three proclaimed it was the environment. 'I knew I was gay since I was three years old,' and another one replied, 'Look, I didn't know what sex was until I was six or seven, so how could you know you were gay at age three?' So, what's the answer? Is it innate or environmental?"

The person who answered with the 'born-gay' claim responded: "Homosexuality is not genetic. I am not a robot run by my DNA but an intelligent man that evaluates my choices. And I made the choice to be gay. I will tell you the problem with the born-gay idea. If we get the label that it is genetic, that implies something biological is wrong with us. That's what Nazi Germany did. They claimed that homosexuality is genetic, along with about every other negative human trait. Then they sterilized us, put us in concentration camps, and murdered us! The experiments which found evidence that part of the hypothalamus is smaller in homosexual men, and is about the size of that in women, caused some problems. One anatomy professor implied that we have brain damage! I have a big problem with the claim that 'I could not help it and was born that way!' No thanks, this explanation is not for me."

The following case is all too common: "I agree with the view that homosexuality is not genetic. One of my best buddies in high school married and, after having five kids, left his wife and moved in with his male lover. It sure hurt his wife! She was stunned, actually devastated, as were all of us. We lived in a small town, so the talk flowed like water when word got out about this scandal."

"The literature indicates that most males' sexual identity is often established by junior high school or at least in high school." He, as his wife told me, had normal sex with his wife for thirty years without a problem and, all of a sudden, he woke up and discovered he had no interest in women?" It's my guess that he must have been bisexual."

Case history: "A guy who abandons his family and wife to move in with his male lover does not have his priorities straight. My wife was paralyzed from the waist down a little over a year after our wedding. Our car was hit by a drunk driver on her side of the car, and I have stuck with her for twenty-two years. We have adopted my brother's two kids. Both my brother and his wife were killed in the same accident. I could have dumped her but, frankly, I would be a jerk to do so. Nor do I fool around, even though I have plenty of opportunities. I am faithful to my wife. In my wedding vows, I promised to care for my wife in sickness and in health, and I am a man of my word. The easiest thing for me would be to divorce her and find some cute young thing, but unlike some people, I have a conscience.

What would people say if he dumped his wife and kids and shacked up with some eighteen-year-old floozy? Isn't that just as bad?"

PUTTING SEX ABOVE ALL ELSE

Case history: "I am fourteen now and want to relate an event that devastated my family. Four years ago, my dad announced that he was gay and was going to move in with his lover. We thought it was a sick joke. He then added, 'I am serious,' and packed his things and left. My mom, my two sisters, and my brother were devastated. Then we became angry. Very angry. He made his choice, his sex life or his family, and he chose his sex life. To me, that is very selfish." *"But he is still your father, and you can still do a lot of things together,"* I added.

"No, my dad is gone, and he is never coming back. He made his choice. Mom is devastated and has not dealt with this very well at all. He tried to visit us a few times, but we are having a hard time dealing with what he did. We also have money problems. The child support he pays does not begin to pay the bills, and Mom had to go to work full-time. There is just so much anger on the part of all of us. His behavior has made it clear where we stand in his life, and we do not want to see him again. He is gone and is not coming back ever."

I thought about this call for several days. His emotional, young voice said much more than what his words attempted to convey.

Another view on this situation: "A guy down the street dumped his wife and several kids, then moved in with some guy, and the neighborhood was very supportive of him. They felt it wasn't right for a gay man to be in a sham marriage and applauded his coming out. I agree with this view. He is gay and should be with a guy, not a woman. I also have a problem with the guy who dumped his wife of twenty years. So, he lived for twenty or thirty years, dating women as an adult, eventually marrying a woman, then one day declared, 'Hey, I am not attracted to gals, but rather guys! I guess I will have to switch partners!'"

"If a guy in his forties with five kids ran off with some eighteen-year-old floozy, we would not say, 'He really likes young girls, so should dump his older woman because that's who he really is.' All decent people would condemn him in the strongest of terms. I have heard men claim they are depressed with their older wife and are very happy when they shack up with some young gal or guy. Well, I get depressed too, and I am sure that, as a 48-year-old man, shacking up with some young, cute, sexy, vivacious, 18-year-old girl would sure lift my spirits! But I have family responsibilities, and I just have to deal with my depression. We have to be responsible as a people for our nation to survive. We all get depressed at times."

Even for the individual who seeks reparative therapy to 'change' sexual orientation (and it works for some but doesn't for others), it is apparently quite a challenge to de-program oneself and requires a lot of effort. There aren't any easy 'solutions,' but it is an ongoing effort to produce good from their situation, whether it's treating AIDS victims or helping a kid adjust to the effects of unstable and crippling family life. What is clear is that if we have an obvious purpose in life, our happiness – and that of others – is inescapably tied to our behavior. Life has an ultimate and good purpose, and part of it is not condemning people caught up in a homosexual lifestyle, claiming that God hates them and that they are going to hell. God IS love and is behind all of the love we see in the world, including a rabbit's love of carrots. Love is experienced as desire.

Incidentally, in over twenty years of attending weekly mass, I have never even heard homosexuality mentioned from the pulpit. Except for a few overzealous individuals, who are missing the spirit of Christianity while maintaining the rhetoric, there isn't a horde of salivating believers out there condemning anyone. On the other hand, I do not support homosexual marriage, even though I do know a few gay couples who obviously care for each other and want to make it public for reasons detailed in the Swedish study.

CHAPTER 5

An Interview with Dr. Beacon

Dr. Richard Beacon is a semi-retired licensed psychologist who has practiced in Michigan for most of his career.

My first question was, *"Could you tell us a little about yourself?"*

Dr. Beacon: "I was born in 1938 in Detroit, Michigan, and attended the University of Michigan for my undergraduate and graduate work, all in psychology. I practiced in Birmingham, Michigan, and in several other cities near Detroit. Usually, I worked for a large clinic as a staff therapist."

"Could you tell us about your early experiences with homosexuality?"

Dr. Beacon: "When I was a teenager, I was strongly sexually attracted to males. Any good-looking young male elicited sexual feelings in me. Several of my psychology professors at the U of M, as well as at least three textbooks that I can remember, taught that the male bonding stage is a normal part of development that almost all adolescents pass through. I was taught that the reason some persons become homosexuals as adults was because they were never able to mature beyond this normal developmental stage. For this reason, my feelings did not bother me a great deal. I felt that they were normal, and I would, in time, simply grow beyond them. I dated somewhat while in high school and in college, although I never became serious with a girl."

"You say your feelings for women were weaker than for men?"

Dr. Beacon: "Yes. My feelings for men were definitely stronger, and my romantic involvements with women occurred largely because this was the role I felt my family and society expected of me. Another important influence in my life was religion. I was never very religious, but my family and many of my friends were. The consistent message I got about homosexuality was that it was revolting, immoral behavior.

Actually, I don't remember any specific incident where my family pointedly discussed this issue. It was just an attitude that I picked up, although occasionally, brief comments were made about this behavior being abnormal or gross. Back in the late fifties and sixties, this message was also consistently conveyed by society. For this reason, I never questioned the idea that homosexual behavior was abnormal."

"Did you overcome your homosexual drive?"

Dr. Beacon: "As time went on, my homosexual feelings became less and less important. I never dwelt on these feelings, nor did I allow them to develop. This was a heck of a lot easier back then compared to today. When I would see an attractive male and have sexual feelings for him, I would tell myself that I must move on to think about other things. Typically, in time, the feelings became increasingly fleeting. They would surface, and I would resolve to put them out of my mind and go on with my life.

It never was a big deal, although it did bother me that these feelings not uncommonly surfaced, albeit usually only briefly. I never thought of myself as a homosexual but simply a person who occasionally had these thoughts. They were worse when I did not have a girlfriend. If I were born today, I probably would be an active homosexual."

"Did you have any attractions to women?"

Dr. Beacon: "I always liked women as people. I actually seemed to get along better with women than with men because they typically had interests that were much closer to mine. My main interest has always been good conversation, reading, interacting with people, and academic subjects like English, history, and psychology.

I never liked sports, working on cars, science, math, wood shop, or any of the other things that other men were supposed to like. I am very much an indoors person and do not even enjoy watching sports, let alone participating. Sexually, I always assumed I was heterosexual in spite of my erotic feelings toward males. My guess is that the closeness and intimacy of another person about your same age can produce erotic feelings in most everybody. Men may adamantly deny this, but men in situations where there are few females – such as in prisons, the military, or other places – often quickly respond sexually to their own sex."

"If you had been born only a decade ago, how would you have dealt with your feelings?"

Dr. Beacon: "I think that if I were a teenager today, I would be far more apt to be labeled a homosexual and end up in that lifestyle. Or I might be bisexual, a label that I never heard of until after college. In my generation, there was a great deal of social pressure not to adopt the homosexual lifestyle, let alone announce to the world that you were such. From my experience with patients, reading and talking to students and young people in general about this subject, it appears that there is now a great deal of pressure to, as they say, 'come out' as a homosexual.

"How have things changed since you were a young man?"

Dr. Beacon: "Just the other day, I overheard two students in a class I am teaching at the University mention a friend of theirs who recently came out of the closet. None of these students had accepted the gay label, yet they all gave a hearty 'All right! Wow!' to the news and recounted with pride the other people they knew that came out. It was a status symbol to know someone who was gay, and my students vicariously experienced the rewards of achieving this status. The whole coming-out process is like a religious conversion, like where one claims they are 'born again' or have 'accepted Christ.'"

"What do you mean by that?"

Dr. Beacon: "Once you publicly commit yourself as gay, it is more difficult to go back and rescind (or waffle) on your identity. Like a marriage ceremony, once you announce to the world that you are man and wife, you have publicly committed yourself, and changing is not very easy. Once a person has 'come out,' he or she sees that label as part of their identity and, as a result, that label guides their future behavior. Any fleeting heterosexual feelings are often repressed with the conclusion that these feelings are strongly against who I am as a 'gay man' or a 'lesbian woman.'"

"How did you meet your wife?"

Dr. Beacon: "We met as undergraduates at the University of Michigan, actually in a psychology class. We were both overachievers, so we studied together. When social events came up, we often went together rather than going alone. Neither of us had much time to do many extra activities, and we rarely went to parties. One cannot study all the time, so occasionally we did fun things together, like going to movies. We also often confided in each other and still talk a lot. Communication was not a problem with us at all. We became best friends but were not lovers. After we both graduated, most of our friends were married, so we felt some social pressure to tie the knot. I actually think that my sexual orientation helped our relationship because sex never got in our way. We were best friends for several years before we progressed to a romantic relationship. So many people marry for sex, but I firmly believe that you should be best friends first."

"Have you adjusted sexually in marriage?"

Dr. Beacon: "I think reasonably well, although I must admit my relationship with my wife as a person, a friend, a confidant, and a partner, are far more important than our sexual relationship. We probably have sex less often compared to most couples, but then we are both in our early seventies now, and we are both still very active in our careers. My wife is an attorney and can easily work twelve hours a day. I think far too much attention is paid to the sex aspect of life in our society today.

Conversely, I have little trouble achieving orgasm and might even be termed as having premature orgasm. This seems to put the blame on the male, when the problem is that it takes longer for a woman to respond than men, and thus I prefer to call the problem 'the need to achieve more simultaneous sexual satisfaction' as opposed to labeling the partner that responds faster 'premature' and the partner that responds slower as 'normal.'"

"Do you think you would have been happier if you had adopted the gay lifestyle?"

Dr. Beacon: "Heavens no!" And I am judging this not only on the basis of my experiences with patients, but also with friends. I have a gay friend in Chicago who is trying to deal with the problem that his parents and almost all of his relatives have now died. All he has left is an uncle, and he is very concerned about being alone when his uncle dies. I have heard this same complaint over and over from gays. The simple fact is that gay relationships often do not last very long, especially between gay males.

Another problem is that those living with partners often have open sexual relationships. Open marriage, they call it. My guess is, if gay marriage were legalized, the divorce rate among gays would be twice as high as that for heterosexual couples, and that a gay marriage would last about half as long, or less. [*He turned out to be correct on this point.*] The research I have read on this point conforms to my own experience. Part of the problem is the high level of promiscuity among homosexual couples. They tend to be more sexually involved than heterosexual couples when the relationship begins but become bored sooner and move on to other partners, even if they stay in the relationship.

Although both my family and my wife's family have passed away, we have many cousins, especially on my wife's side (she came from a large family). Besides, we have three children and seven grandchildren, all of whom we are fairly close to, so I don't have the same worries as my friends who have adopted the gay lifestyle. I can't ever imagine being without family. If I adopted the gay lifestyle, this part of my life would not have existed."

"Why is that?"

Dr. Beacon: "First of all, there is the low life expectancy problem among gays because of disease, depression, and related problems. Practicing therapy for almost forty years has allowed me to work with a large number of patients, and I can generalize from my own experience. Those living the homosexual lifestyle, including females (which surprised me), are simply far more likely to develop some sexually transmitted disease, or become sick more often, and they also die much younger. Most of my homosexual patients died in their forties, and, although many died of AIDS, many died of other diseases common to homosexuals.

I have had only a few gay patients live to be elderly. Of these who lived to be old, even though they saw themselves as gay, they tended to have very few long-term sexual relationships for a variety of reasons. Some were just extremely busy in their career and did not want to get entangled in sex that could cause problems. Others did not seem to have a high sex drive anyway and usually did not miss sex. For them, the gay lifestyle was an important means of making social contacts, and this seemed to be more important than sex for them. Yet, others, even though they had a long-term relationship with a person of the same sex, did not find sex all that important. This was especially true of females."

"Have your patients influenced you in some way?"

Dr. Beacon: "Although I am making these judgments largely on the basis of my patients, nonetheless, important insight can be obtained from clients because they are more likely to be honest in a therapeutic relationship than in a research interview. If they want help with a mental problem, they often realize they have to open up with their feelings. Most of my patients appear to freely open up. From my own experience, I can see very good reasons to discourage people from involving themselves in the gay lifestyle."

"Do you think that gays can change?"

Dr. Beacon: "The idea that people cannot alter their sexual orientation is often untrue. People can, and do, modify their sexual practices as well as other practices. I have asked most of my patients involved in the gay lifestyle if they felt they could be heterosexually active, and many said yes. Only a few could not. I think a major part of the problem is labeling. Many recognize this and label themselves bisexual. One patient's girlfriend claimed that her boyfriend was 'forty percent gay.' They eventually split up after college, and he now works for a gay center helping others on the fence to move into the gay lifestyle."

"Why is that?"

Dr. Beacon: "To some degree, becoming gay in society today is a fad that I think will have horrendous social implications in the future. I really believe our society is headed down the wrong course, and I think telling people that they should come out of the closet and have a good time living the gay lifestyle may seem like a short-term solution, but, overall, it will be very harmful to society. Of course, there are exceptions, but, as a whole, I perceive that it is not a good thing."

"How do you deal with the problem of people saying you are repressing your sexuality and becoming something that you are not born to be?"

Dr. Beacon: "I don't think this is a valid comparison because it assumes that our sex drive is rigid and fixed, when it is actually quite flexible. People develop all kinds of fetishes and sexual interests, some of which are harmful and should be repressed. For example, I have a number of patients who are pedophilia-oriented. One male in particular had a very strong erotic attraction to young teenage boys. Obviously, acting on this attraction could cause him serious problems. After he married, he was still bothered by this attraction, so he sought counseling. This is how I met him. His marital adjustment was actually fairly good. Pedophiles are not void of other sexual desires, but, at most, their pedophilia is a strong or even powerful focus of their sexual energies. As far as I am aware, he has managed to control this impulse and has had a normal sexual adjustment."

"Could you explain?"

Dr. Beacon: "He realizes that this sexual outlet is forbidden and could cause him a great deal of difficulty in his life. He has accepted that this sexual outlet is unable to be fulfilled. Although he admits that the feelings are still there, he has moved on with his life. He has made at least an average sexual adjustment with his wife. They have three children and have been married for close to thirty years now. It was my recommendation in his case that he not discuss this issue with his wife, as this could cause major problems. Some might disagree with me, but, in my experience, this can be disastrous."

"Very interesting. I tend to agree with you here."

Dr. Beacon: "I had another patient who revealed his feelings to his wife, and it ended their marriage, even though his marriage was probably above average and he had achieved a very good sexual adjustment. From that point on, his wife saw him as a pervert and wanted nothing to do with him. The problem is that many people believe that one either has pedophilia drives or one does not, and there is no in-between. The fact is, many of these feelings are fairly common, but most of us can effectively control them. This may explain the enormous popularity of magazines such as one titled Girls Just Legal, or something like that. The claim is made that the girls who pose in it are eighteen or over, but look fourteen or younger, and, in essence, magazines like this are legal child pornography."

"Is the same thing true for homosexuality?"

Dr. Beacon: "Yes. Since a large number of homosexuals are promiscuous, several steps are involved in dealing with this problem. The first is to cut back on the promiscuity, then to move from homosexual to heterosexual relationships. The longer one has been in homosexual relationships, the more difficult this move typically is. For people who never became deeply involved, it's not that difficult, at least for most, to leave this lifestyle. Of course, for some people it is, but in my experience, with help and support, most can adjust fairly well. Today, due to much more social approval of the gay lifestyle, it has become far more difficult to change."

"Thank you for your time and insight!"

Dr. Beacon: "You're very welcome."

CHAPTER 6

The Bible's Teaching

From the Biblical and the classical Christian perspective, sex is a biological drive designed to fulfill the command to "be fruitful, and multiply, and fill the earth," and produce a monogamous bond to care for the couple's children and for each other (Genesis 1:28; 2:18-24; 9:1-7 KJV). One of the many passages that are cited to condemn both male and female homosexuality is Romans 1:18-28 (KJV), which reads, "For this reason God gave them up to dishonorable passions. For their women exchanged natural relations for those that are contrary to nature; and the men likewise gave up natural relations with women and were consumed with passion for one another, men committing shameless acts with other men and receiving in themselves the due penalty for their error. And since they did not see fit to acknowledge God, God gave them up to a debased mind to do what should never be done." This scripture concludes that persons who committed homosexual acts will receive "*the due penalty for their perversion.*"[1] This argues that part of this "penalty" is the health consequences that result from homosexual behavior. This view is widely regarded as the reason for many of the Levitical laws.[2]

1 Timothy 1:9-11 adds that "men who lie with males" (some translations use the word sodomites, and others, such as the *New American Standard Bible*, use homosexuals) are engaging in acts "in opposition to the healthful teaching according to the glorious good news of the happy God."[3] Notice the term *healthful*

---

1 New International Version (NIV); emphasis mine.

2 Thomsen, Russel J. 1974. *The Bible Book of Medical Wisdom*. Old Tappan, NJ: Fleming H. Revell.

3 *Interlinear Translation of the Greek Scriptures*. 1985. Brooklyn, NY: International Bible Students Association. Also see Leviticus 10:11-15, 18:22, and 20:13; Deuteronomy 23:17; I Kings 14:24.

in this translation. *Strong's Concordance* states that the Greek word used here (#5198 *ugiainoush*) means healthful, and for this reason, many modern translations use the term "healthful."

Other translations use the word "sound" instead of "healthful," which would not contradict the meaning emphasized here. Jude 7 adds that after the people of Sodom and Gomorrah, who lusted "after flesh for unnatural use, are placed before [us] as a [warning] example by undergoing the judicial punishment."[4] The term "Sodom" is the basis for the word "sodomy," which is used to designate homosexual behavior (cf. Genesis 19:5-7, 24, 25).

The seriousness of homosexual behavior is indicated by Leviticus 20:13, which mandates the death penalty. Strong condemnation is also taught in the New Testament. For example, 1 Corinthians 6:9 states that "none who are guilty of either adultery or of homosexual perversion" will inherit God's kingdom (*New English Bible*). Note that the scriptures label these acts as behavior that can be changed: "and such were some of you" (1 Corinthians 6:11). The scriptures also teach that sexual immorality emerges from the human sinful nature, not biology (Galatians 5:19). It also teaches the human body is not constructed for sexual immorality (1 Corinthians 6:13; see also 1 Timothy 1:10; Matthew 15:19; and Romans 1:26-27). Beck translates 1 Corinthians 6:13 as follows: "the body was not made for sexual sin."

These texts and others, plus the extensive writings of the so-called "church fathers," have been the major historical basis for the Christian condemnation of homosexual behavior.[5] The Jewish and Muslim positions have been both historically similar to the Christian view and were both based on similar reasoning.

In a survey of church positions on homosexuality, professor, minister, and New Testament scholar Jeffrey Siker notes that most church policy statements "... consider homosexual orientation as a distortion of God's design and homosexual behavior as sin" (1994). Ellis and Ames note that "in the Western world, heterosexuality was attributable to what God had ordained as natural and good," and all deviations from it were viewed as harmful (1987, p. 233). This position has been the majority view in the West for most of the last several millennia.

The Scriptures teach that God's laws were given for the ultimate *benefit* of humanity, which is why God could promise the Israelites that if they followed these laws, they would have "none of these diseases" (Exodus 15:26). Colson and Pearcey (1999, pp. 15-16) conclude that "God created our bodies and the moral laws that keep us healthy" and that "If we want to live healthy, well-balanced lives, we had better know the laws and ordinances by which God has structured creation.

---

4 Interlinear Translation of the Greek Scriptures, 1985.

5 See McNeill, 1976 and Soards, 1995 listed in the bibliography in the end of this book.

This understanding of life's laws is what Scripture calls wisdom." This summary argues that a major reason *why* Scripture condemns homosexuality is that it has clear adverse effects on human health. Many Biblical prohibitions (such as the quarantine laws and certain others) were plainly given for health reasons.[6]

THE EFFECT OF HOMOSEXUAL BEHAVIOR ON HEALTH

A review of the medical literature shows that homosexual behavior clearly has a major detrimental effect on health.[7] The adverse effects of homosexual behavior on health help to explain why studies found that the median lifespan for the average male involved in a lifelong homosexual lifestyle was in his 50s, while females averaged a few years longer (corresponding to an over 30-year lifespan decrease).[8] The major reason for early death is due to the transmission of numerous contagious and potentially lethal diseases through the practice of sodomy and other common homosexual behaviors.

Until recently, autoimmune deficiency disorder (AIDS) alone reduced an active homosexual's life expectancy by about ten percent. A lot of despair, illness, and death, are all a common part of the gay lifestyle. Rampant promiscuity is common, perhaps because of the inherent inability of homosexual behavior to sexually satisfy. For this reason, sexual-enhancing drugs tend to be commonly used. "Gays are promiscuous men who worship youth and are often after sex and nothing else. One older gay man was bitter, stating, 'I was very popular as a strong, in-shape, good-looking youth, but hardly any men want me now that I am older and have lost my youth'."

One case history illustrates this problem: "I am a straight guy and grew up in a big city. My best friend since preschool and I were really close until he passed away. He was gay, although the gay idea didn't surface in him until junior high. He and I hung around with a bunch of other guys, eight of whom came out as gay

---

6 See Greenblatt, 1963; Thomsen, 1974; and McMillen and Stern, 2000 in the end of the book.

7 Abraham, Jerrold L. 1980. "Medical aspects of homosexuality." *New England Journal of Medicine* **302**(8):463-464, February 21; Byne, 1994; Penn, Robert. 1997. *The Gay Men's Wellness Guide.* New York, NY: Henry Holt; McMillen, S.I., and Donald E. Stern. 2000. *None of These Diseases: The Bible's Health Secrets for the 21st Century.* Grand Rapids, MI: Fleming H. Revell.

8 Cameron, Paul, William Playfair, and Stephen Wellum. 1994. "The longevity of homosexuals: Before and after the aids epidemic." *Omega* **29**(3):249-272, November; Cameron, Paul, Kirk Cameron, and William Playfair. 1998. "Does homosexual activity shorten life?" *Psychological Reports* **83**(3 Pt. 1):847-866, December; Cameron, Paul. 2002. "Homosexual partnerships and homosexual longevity: A replication." *Psychological Reports* **91**(2):671-678, October.

while still in high school. We all went to a big high school, so eight gays were not all that many out of the four thousand or so attending the school. I am fifty-four now, and they are all dead. My best buddy died at age thirty-two, and the last of the eight died at the young age of forty-nine. Three died of AIDS, but several others died of complications from other STD-related diseases [STD = sexually transmitted disease]. One good friend died of penile cancer and another of anal cancer. It was not easy watching my best friends slowly die, often suffering greatly. It may be unusual, but that's my story. I think a lot more effort needs to be expended on preventing diseases in the gay world. It's a big problem in their culture, and hardly any of my heterosexual friends have died."[9]

Sex can, and commonly does, cause disease in both heterosexuals and homosexuals, but this topic is especially a concern for the homosexual community. Another study found the median age of death for partnered men was 45, and for non-partnered men was 46, indicating partnering was not effective in decreasing the health risk.[10] Violence within the homosexual community is another major problem that helps to explain the low level of life expectancy.[11]

The phrase "involved in homosexual behavior" is used because the concern here is with *behavior*, whether engaged in by those who consider themselves heterosexual, bisexual, or homosexual. Sexual *behavior* produces a high risk of contracting STDs, not sexual orientation.[12] Homosexual behavior is largely a learned behavior that is influenced, as are all human traits, by genetic factors.[13] Adolescent physician Meeker noted that:

> Regardless of the reason for its happening, homosexual activity, particularly between boys, is extremely dangerous, especially if they engage in anal intercourse. The anus opens into the rectum—the lower end of the large intestine—which is not as well suited for penile penetration as the female vagina is. Both the anus and rectum have rich blood supplies, and their walls, thinner than the walls of the vagina, are easily damaged. When penetration occurs, it's easier to tear blood vessels, which in turn increases the risk of acquiring or receiving an infection as

---

9 Grossman, Lisa. 2015. "Time to lay 'born this way' to rest." *New Scientist* **227**(3031):18-19, July 25.

10 See Cameron, 2002 in the biography in the end of the book.

11 See Burke and Follingstad, 1999; Cameron, et al. 1994 in the biography in the end of the book

12 See Chrestiansen and Lowhagen, 2000 in the biography in the end of the book.

13 Santinover, 1996; Bagemihl, Bruce. 1999. *Biological exuberance: Animal homosexuality and natural diversity*. New York, NY: St. Martin's Press.

penile skin and/or semen come in contact with the partner's blood or semen.[14]

Sexually transmitted diseases (STDs) usually result from intimate sexual contact but can also be transmitted by other means. STDs are caused by several dozen types of bacterial, viral, parasitic, and fungal infections. Women disproportionately suffer from their effects partially because women are less able to protect themselves from exposure to STDs. Furthermore, the long-term complications from STDs in women are more likely to be more severe than in men, including infertility, ectopic pregnancy, and cancer.

Case history: "I was diagnosed with anal cancer a few months ago. The doctor asked me if I was gay. I replied, 'No, I am married, why do you ask?' He said that gay men are seventeen times more likely to develop anal cancer than straight men. I was homosexually raped a few years ago in prison, so I asked him if that may have been the cause. He replied, 'Could be, but it is hard to say for sure at this late date.' He then found out my cancer was the Human Papilloma Virus-related type, so he concluded that the rape was likely the cause. Is this a common problem? 'Rape is a common cause of anal cancer, as well as sex with someone who has the HPV. I also know that in prisons, homosexual rape of heterosexuals is a problem. If you call an inmate 'gay,' they will at times become very angry, yet will indulge in homosexual sex.'"

Furthermore, the enormous success in treating some STDs, such as syphilis, has led to a recent decline in vigilance and concern about this problem. By the mid-1980s, though, *Chlamydia trachomatis*, *Trichomonas vaginalis*, herpes simplex virus, the human papilloma virus, and one of the most important, HIV [Human Immunodeficiency Virus], now dominate the field of STDs.

For many diseases, such as HIV, the transmissibility is much higher for penile-anal intercourse than for penile-vaginal intercourse (Scutchfield and Keck, 1997, p. 291). A major reason for the continual transmission of many STDs is a core population of highly sexually active persons who serve to spread it to many different people. At the very least, lowering the number of sex partners could also be an important means of reducing the problem. The reproduction rate of bacteria and viruses must be greater than one, and if it is less than one, the disease will soon die out.

A high percentage of men involved in homosexuality engage in extremely risky behavior that puts them at a high risk for not only AIDS, but also many other STDs.[15] Homosexual behavior produces a venereal disease rate as much as 22

14 Meeker, 2002, p. 152

15 McKusick et al., 1985; Lemp et al., 1995; Elford et al., 1999; Stephenson, 2000)

times above the national average. Not only do gonadal-anal sexual practices produce major health risks among homosexuals, but the sexual practices in which homosexuals engage (a major one for males being anal intercourse) carries major health risks, including not only AIDS but also non-STD infectious diseases such as tuberculosis, and virus-causing warts which can spread rapidly and cause secondary infections, abnormal hemorrhaging, and even many types of cancer.[16]

Many diseases are spread by body wastes, especially fecal matter, which historically has been directly or indirectly the major cause of the spread of most major infectious diseases. These diseases are also effectively spread by homosexual behavior, and only a few are covered here. Parasitic amebiasis and urethritis, viral herpes, pediculosis infestation, condyloma, amoebic colon infections, and anal and penile cancer are all common in the homosexual population.[17]

16 Beral et al., 1992 ; Chu et al., 1992; Dooley et al., 1992; Koblin et al., 1996; Frisch et al., 1997

17 See Rueda, 1982, pp. 52-53; Palefsky et al., 1998.

CHAPTER 7

The Case History of Edmund White III

A *New York Times* obituary of Edmund White III, a man driven by sex who claimed he had had thousands of homosexual lovers, authored by Fred A. Bernstein in his June 4, 2025, article, lauded White's long life and success as a multiple best-selling fiction and nonfiction author. White's writing glorified his promiscuous life with many, often well-known, men. His death was confirmed by his husband, Michael Carroll, who said White, a morbidly obese man, collapsed while weakened by "a vicious stomach bug," adding that the cause of his death was unknown. The obituary article, titled "Edmund White, Pioneer of Gay Literature, Is Dead at 85," described his many honors which included "his vast and varied catalog of sexual experiences, in more than 30 books of fiction and nonfiction and hundreds of articles and essays, becoming a grandee of New York literary life for more than half a century." Although raised in the Church of Christ, he rejected Christianity and later identified as an atheist.

White was given countless major literary awards for his writing on his sexual escapades. He taught writing at several Ivy League universities, including Brown and Princeton, where he was on the faculty from 1999 to 2018. One of his highly acclaimed books chronicles his first 65 years, with chapter titles including "My Shrinks," "My Hustlers," and "My Blonds." His frequent companion was the young French architect, Hubert Sorin, whom White called "the love of my life." Sorin died at age 32 in 1994 in Marrakech, Morocco, from AIDS-related complications. As the partner of American author Edmund White, he collaborated on the 1994 book *Our Paris: Sketches from Memory*.

White was born in Cincinnati, the second child of school psychologist Delilah White and chemical engineer "famous womanizer" Edmund Valentine White, II. His mother was a school psychologist who, in Mr. White's telling, practiced on her son, administering a series of Rorschach tests and diagnosed him as "borderline psychotic." When White was seven, his father left his mother for a younger woman. As an adolescent, a psychiatrist labeled White "unsalvageable."

Although accepted to Harvard, his Detroit psychiatrist insisted that he continue psychiatric treatment with him. So, he attended the University of Michigan, graduating in 1962. White then moved to New York to work for Time-Life Books while writing his own books at night. In 1970, he worked as an editor for the *Saturday Review*.

He often found sex by cruising the streets, but told *T* Magazine in 2024, "to make myself stay in and write, I would hire hustlers." He observed, "We had all thought that homosexuality was a medical term. When he saw that we could be a minority group — with rights, a culture, and an agenda," he began working to expand the movement. In 1995, he began a sexual relationship with Michael Carroll, a writer 25 years his junior. White, who was married then, had no intention of being monogamous.

In his 60s, after a sadomasochistic relationship with an actor half his age ended, he was diagnosed HIV positive but four of the seven male members of his gay writers group succumbed to the disease, as did his two closest friends, literary critic David Kalstone and Bill Whitehead (White's editor at Dutton, his publishing group), and scores of other friends and lovers. In 2000, White told *The Guardian* (a British daily newspaper) about his guilt in living to an old age while so many gay men he knew died young.

My thought after reading this article was, "Why was a man who lived a life filled with promiscuity and deception highly rewarded by academia and the major media? Do we really want young people to look up to and imitate this man? Men who live virtuous lives and love their wives and kids should be upheld as examples, not men who live a life of total depravity contrary to Judeo-Christian values.

CHAPTER 8

Sexually Transmitted Diseases and Homosexuality

The major STD of concern today is AIDS. Although in the West, homosexual sex is a leading cause of the AIDS problem, the problem has been exacerbated by the spread of AIDS from the homosexual to the heterosexual population. This is due to many factors, such as the fact that many persons in the homosexual community do not restrict themselves to strictly homosexual behavior.[1] AIDS is especially difficult to control because many infected people do not develop clear symptoms of the disease until about 5 to 10 years after they contract the human immunodeficiency virus (HIV).

For decades, about 60 thousand *new* HIV cases were reported each year in America alone. By late 1999, over 700,000 cases of AIDS had been reported in America, and over 30 million in the world (Chin, 2000; *Statistical Abstract of the United States*, 2000). Penn (1997) claims that fully one in two sexually active, promiscuous, homosexual men are HIV-positive, and close to half of all new HIV infections are in this population.[2] Since 1999 alone, a 17 percent increase in HIV diagnoses has occurred among gay and bisexual men (CDC data).

The CDC also reported that 39% of homosexual and bisexual men interviewed admitted to having unprotected sex with someone whom they met over the internet. Furthermore, the AIDS epidemic was growing rapidly internationally, especially in many third-world nations, until recently.[3] The problem is now a pandemic, and is

1 Melbye and Biggar, 1992.

2 Wolitski et al., 2001.

3 Cock and Weiss, 2000

most serious in Africa, where up to half of the teenagers living in some countries are expected to die of AIDS.[4] So far in the USA, AIDS has taken an estimated 260,000 lives of men who have had sex with other men (MSM).[5]

Not only do their gonadal sexual practices produce a major health risk among homosexuals, but *human herpes virus 8*, the cause of Kaposi's sarcoma, is also epidemic among homosexuals–primarily due to oral sexual transmission of infectious saliva (Lorber, 1996; Pauk et al., 2000). In Kampala, Uganda, Kaposi's sarcoma associated with AIDS is now the most common type of cancer in both sexes.[6] Persons who have homosexual sex are 20 times more likely to develop Kaposi's sarcoma than AIDS heterosexual patients. Herpes virus, including herpes simplex, which is also incurable, is now epidemic in the gay community (Penn, 1997). Herpes can be extremely painful and often leads to other serious medical complications.[7]

GONORRHEA

Gonorrhea is now considered epidemic in the gay population. This disease is a major problem for many reasons, especially because it can foster transmission of the AIDS virus by weakening the body's defense system, and the bacteria act as a HIV carrier (Munoz-Perez, 1998; Fox et al., 2001).

Gonorrhea is the most prevalent venereal disease in history and has been known since Biblical times. Gonorrhea is also the most common STD worldwide, and the most commonly reported communicable disease in the U.S. More than one million new cases of gonorrhea are reported annually, and unreported cases are probably equal to this rate. In homosexuals, it commonly infects many different parts of the body, including the urethra (the tube in the penis), anus, mouth, throat, or vagina. Gonorrhea has almost tripled among the homosexual population from 1992 to 1999 (from 5 percent to 13 percent of all samples), and rectal gonorrhea is now a major problem in this population (Hegyi et al., 1997; McMillan, Young, and Moyes, 2000; Wolitski et al., 2001).

Gonorrhea infects only humans and is a sexually transmitted bacterial infection caused by *Neisseria gonorrhoeae*. In males, the bacteria often infect the urethra (the tube inside the penis that carries urine or semen). In women, it can infect the vagina, uterus, and fallopian tubes (Singh and Mohanty, 1999). All sexually active persons, especially promiscuous gays, are at risk (Goldstone, 2001). Symptoms usually begin two to five days after infection.

---

4 Cock and Weiss, 2000; Whyte, 2000.

5 Wolitski et al., 2001.

6 Wabinga et al., 2000

7 McMillen and Stern, 2000.

Gonorrhea is spread via anal, oral, or vaginal sexual contact with an infected person. The only sure means of prevention is abstinence or monogamous sex with an uninfected partner. During sex, the discharge can seep into the anus, urethra, or vagina. Gonorrhea is easily passed on during sex with an infected partner. Although penetration is not necessary, oral and anal sex are frequent ways gonorrhea is passed from one person to another partner. Gonorrheal infection also makes it much easier for both persons to become infected and transmit HIV (Goldstone, 2001).

In men, the most common symptom is a greenish-yellow, penile discharge. Although the discharge is usually profuse, some men notice only a stain on their underwear. The discharge is often accompanied by dysuria (a burning pain when urinating). In active homosexuals, the common early symptoms include a sore throat or soreness in the anal area. Doctors often do not think to culture these areas of the body unless they know the patient engages in anal or oral sex (Goldstone, 2001).

In women, gonorrhea infects the vagina and quickly spreads into the uterus and fallopian tubes. Symptoms include severe lower abdominal and pelvic pain, accompanied by fever and nausea, a condition known as pelvic inflammatory disease (PID).

Gonorrhea is often diagnosed by culturing body mucous, or other discharges. If the discharge is minimal, a doctor often puts a tiny swab into the urethra to obtain a sample. Because the bacteria are extremely sensitive to dry environments and require a high concentration of carbon dioxide to grow, culturing requires very strict conditions. Because of the false negative problem, doctors frequently send a swab sample to a laboratory for DNA analysis (Goldstone, 2001).

When gonorrhea infects the mouth, throat, or anus, the diagnosis is usually more difficult. Gonorrhea of the mouth, throat, and anus (common locations in gays) is often difficult to diagnose because the symptoms are often atypical and cause less pain and discharge than during penile or vaginal infections (Ahmad and Sukthankar, 1998). Throat and mouth infections resemble a typical "sore throat." The pain may be minimal and often goes away fairly quickly after the infection. Gonorrhea in the anus often causes anal discharges or bloody bowel movement, and severe pain is often present.

SYPHILIS

Syphilis is a highly contagious disease caused by *Treponema pallidum*. The bacteria can live only for a few minutes outside of the body. Therefore, they are normally transferred by sexual contact, especially anal intercourse. Transfer is most effective if a break exists in a mucous membrane, as is common in anal sex.

The Centers for Disease Control in Atlanta, Georgia, estimates that over 40,000 syphilitic infections occurred in 1999, at least *half* of which were in men who had sex with other men (most studies indicate gays are about 2% of the population) (Communicable Disease Report, 2000; Bernstein et al., 2001). The CDC noted that syphilis is on the rise in the U.S. "largely because of outbreaks among gay and bisexual men" (CDC report, 2002, p. 12). If untreated or not treated properly, it can result in widespread damage to the central nervous system, heart, and aorta, often causing heart failure, general paresis (impaired movement), insanity, and, if not treated, death.

ARE CONDOMS THE SOLUTION?

Condoms are only partially effective (or impractical) for the major types of sex in which homosexuals commonly engage, and many dislike using condoms even for behaviors for which they are appropriate (Van de Ven et al., 1997; Stephenson, 2000). Many people do not use condoms for oral sex or during foreplay, and the failure rate is estimated to be from 5 to 8 percent (Darrow, 1989; Carey et al, 1992; Duyves, 1993; d'Oro et al., 1994; Faundes, 1994; Donovan, 1995, 1995a; Feldblum et al., 1995; Bounds, 1997; Spruyt et al., 1998; Deparis et al. 1999; Macaluso et al., 1999; Mekonen et al., 1999; Sparrow et al., 1999; Stone et al., 1999; Taylor et al., 1999; Feldblum et al., 2000; Wong et al., 2000; Feldblum et al., 2001; Golombok et al., 2001).

Other disease problems common among those involved in the homosexual lifestyle also pose a serious health risk, such as what is called "water sports," which involve urination on one's partner; and "fisting," the insertion of the fist in one's partner's rectum, a practice that often causes rectum and sigmoid-colon tearing and infections that can lead to septicemia and severe complications, including death, in immune compromised persons (Day, 1991).

TRADITIONALLY NON-SEXUALLY TRANSMITTED DISEASES AND HOMOSEXUALITY

Many traditionally non-sexually transmitted diseases are also much more common among homosexuals than heterosexuals. For example, sperm easily penetrates the colon wall and, once inside the body, it adversely affects the immune system. This results in a greater vulnerability to a variety of diseases (Biggar et al., 1984; Mavligit et al., 1984; Meeker, 2002).

Non-self-sperm that enters the anus or urethra of a man commonly invades the delicate mucus linings and enters the bloodstream. Sperm and certain semen secretions invading a body are in some ways the immunological equivalent of its reaction to bacteria and viruses, or even the transplantation of cells or organs

(Tyler, 1994). Antibodies attacking the sperm can attack the corresponding cells and organs of the host. The result may cause autoimmune diseases, including arthritis, diabetes, thyroiditis, and lupus erythematosus.

Sperm deposits in the sigmoid colon may even have a role in AIDS, urinary infections, congenital and inherited defects, and atherosclerosis. It may influence the development of many diseases, both local and systemic, including cancer (Tyler, 1994). Disease symptoms are often absent at the sperm entry site, but entry-site diseases can include mucoid and purulent discharges, often similar to gonorrhea, and nonspecific urethritis and sores, which may be similar to herpes, chancroid, or syphilis (Tyler, 1994).

The fact that male and female gonads were designed for each other leads logically to the conclusion that same-sex romantic couples will experience more conflicts. The research confirms this: Burke and Follingstad (1999), in a review of 19 studies, found that a much higher prevalence of abuse exists among both lesbian and gay populations compared to heterosexual populations. Other research has found that homosexuals are more prone to substance abuse, including smoking, higher rates of school-related violence, suicide—often from twice to four times as high—and also depression (see Herrell et al., 1999; Averbach, 2000; van Heeringen and Vincke, 2000; Rohde et al., 2001).

Homosexual behavior also commonly transmits many other non-sexual diseases that are rare among heterosexuals. For example, homosexuals as a group are far more apt to have a wide variety of bowel diseases, which are generally lumped together under the designation "gay bowel syndrome." Many types of infections, such as prostatitis (inflammation of the prostate gland), an often chronic and extremely painful condition, are also common in homosexual men (Penn, 1997).

HEPATITIS B IS NOW AN EPIDEMIC AMONG THE HOMOSEXUAL POPULATION

Much of the sexual behavior common among homosexuals is objectionable from a general health standpoint. Active homosexuals with multiple partners have a rate of infectious disease almost ten times higher than the general population. Both classical venereal diseases and other diseases, such as hepatitis, have now been epidemic among the homosexual population for decades (Christenson et al., 1982; Penn, 1997). Over 300,000 new cases of hepatitis are diagnosed annually in the U.S, and fully 80% of homosexuals have evidence of exposure to the hepatitis virus compared to only 5% of the rest of the population (Clark, 1995, p. 115).

Hepatitis can be contracted from an infected person through anal or vaginal intercourse, shared needles (including tattoo needles), toothbrushes, razors, and sex toys—all that is required is the transfer of a microscopic amount of infected blood or body fluids (Goldstone, 2001). Since the hepatitis B virus is in the blood, semen, and other bodily fluids of infected individuals—people who engage in unsafe anal or vaginal sex, or use needles, toothbrushes, razors, tattoo equipment, or sex toys that contain contaminated blood—all are at risk (Goldstone, 2001).

Hepatitis B is one of three common viral infections of the liver (hepatitis A and hepatitis C are the other two). Hepatitis B, a viral disease several hundred times as infectious as AIDS, induces both the chronic and acute form of hepatitis, either of which can be fatal (Clark, 1995, p. 115). It can lead to cirrhosis of the liver, a severe, irreversible scarring of the liver. Cirrhosis of the liver is a major cause of liver cancer and even death. Hepatitis B infects 140,000 to 320,000 Americans annually, and over 1.25 million persons have chronic hepatitis B infections. Before the HIV epidemic in 1980, hepatitis B was the most dangerous sexually transmitted infection gay men faced (Clark, 1995).

Men who have sex with men are at especially high risk for hepatitis B (Remis et al., 2000). Corey and Holmes (1980) found that the annual incidence of infection of hepatitis A among homosexual men was 22%. In contrast, not a single heterosexual man acquired hepatitis A during the study. The researchers concluded that contracting the virus correlated strongly with homosexual behavior. The infection agent is a *hepadnavirus* that must enter liver cells in order to reproduce (Chin, 2000). The incubation period is usually from 45 to 180 days, depending on the amount of virus in the inoculum, the amount of transmission, and the health of the host (Chin, 2000).

Hepatitis B symptoms are extremely variable, both in severity and duration. Some people become deadly ill, while others do not realize they have been infected for some time. Most people are contagious before they know they are sick, and many patients never knew they had hepatitis until a routine blood test showed them that they have antibodies to the virus. The most common symptom is profound fatigue—sometimes to the extent that the patient is bedridden.

Other common symptoms include muscle aches, nausea, and vomiting (often worsening as the day progresses), loss of appetite, fever, and a dull upper abdominal discomfort. Loss of the desire to smoke is a classic sign of hepatitis. Most people develop jaundice (their skin turns yellow, a trait most noticeable in the whites of their eyes) as a result of the liver's inability to process bile. The urine darkens, and the stool can turn a sandy color.

Hepatitis is confirmed by blood tests, which check for parts of the hepatitis B virus or antibodies that the body manufactures to defend itself. Hepatitis B virus particles and antibodies differ from those found with other types of hepatitis. Most often, the virus infects the liver, and if the patient is healthy enough, the body's own immune response wages a battle that eventually destroys the virus (Goldstone, 2001). In approximately 5 to 10 percent of infections in people with normal immune systems, the immune system cannot clear the hepatitis B virus from the liver, and the liver is gradually destroyed (chronic active hepatitis).

Hepatitis can also lead to cirrhosis of the liver, liver failure, and death. In people with HIV and other immune disorders, hepatitis B becomes chronic almost 90 percent of the time. Although hepatitis is dangerous in any form, one type, called *fulminant hepatitis,* rapidly destroys the liver. People who contract this hepatitis strain often eventually lapse into a coma and die within a week (Goldstone, 2001).

No medications exist that can effectively treat acute hepatitis. Treatment is generally "supportive," the key components being rest, good nutrition, and monitoring to ensure that the liver heals. Because nausea is common with hepatitis, it is therefore best for infected patients to eat their larger meals early because nausea often worsens as the day progresses (Goldstone, 2001). Hospitalization is necessary if one cannot take in adequate food and water or is too weak to remain at home. All alcohol and drugs (which further tax the liver) must be avoided until recovery is complete (which must be determined by the doctor, not the symptoms). The doctor must monitor both liver function and nutritional status.

If the hepatitis B virus remains in the blood for more than six months, the infection is labeled chronic. A liver biopsy is recommended before any further treatment is undertaken. If the biopsy indicates major liver destruction has occurred, interferon injections are given to help cure chronic infection. Interferon therapy, however, carries major risks itself, including severe fevers, shaking chills, and, in many people, bouts of serious depression occur. Interferon works best in patients who are HIV-free (Munoz-Perez, 1998).

Chronic hepatitis can also be treated with the oral anti-retroviral medication, Epivir HBV (lamivudine), which targets an enzyme required for hepatitis B virus reproduction. This drug, though, has many side effects, including headache, nausea, fatigue, diarrhea, and life-threatening inflammation of the pancreas. The side effects depend on the patient's general health and can be very severe.

ANAL SQUAMOUS CELL CARCINOMA, TYPHOID, AND DYSENTERY

A strong relationship exists between homosexual promiscuity and the risk of anal squamous cell carcinoma, which in most cases is caused by a sexually transmitted pathogenic human papilloma virus (Frisch et al., 1997). A history of receptive anal intercourse, a history of sexually transmitted diseases, more than 10 sexual partners, and HIV infection are all common predisposing factors (Ryan et al., 2000).

Many diseases normally caused by drinking contaminated water are also common problems among homosexuals. The CDC has so far identified seven gay men infected with typhoid through anal intercourse or oral-anal sexual contact.

Typhoid, caused by *Salmonella typhi,* is usually transmitted by water and food tainted with human feces (Adams, 2001, p. 20).

The most common cause of chronic dysentery is the amoeba, *Entamoebia hystolytica.* Dysentery occurs when the amoeba is in the reproducing, nondormant trophozoite stage. The cyst stage is dormant, although infective. The primary way people become infected in a modern nation is by ingestion of, or by other contact, such as anal intercourse, with human feces. The disease can be lethal if chronic. In response to disease and the homosexual issue, Fox noted that the

> colon and rectum are *made for* the elimination of fecal matter and not for sexual experience. Fecal matter is eliminated because it is indigestible and contains disease-causing materials. With sexual penetration, the rectal muscles are often torn or over-expanded, and the fragile lining of the colon is almost always torn. The tearing of the colon allows fecal matter to penetrate into the body, bringing with it infectious disease. (Fox, 1994, p. 2; italics in original)

Despite the best health care system in the world, a large majority of those who engage in long-term homosexual behavior in America live close to half the lifespan of the average healthy person (Cameron et al., 1989, 1994). The disease problem in the homosexual community is so common that many gays patronize doctors who specialize in treating homosexuals. Homosexual social networks and the homosexual press are common sources that patients use to contact such physicians.

A problem with repeated infections common in the gay community is that antibiotics can delay death, and even cure many bacterial venereal and other diseases, but they have many side effects. Eventually, resistant strains often emerge that may prove to be lethal and that can be easily communicable to others. Overuse of antibiotics is a major concern, as is the attitude that infection is not a major concern because antibiotics will take care of the problem. This attitude is irresponsible and results in behavior that, in the long run, is potentially lethal.

SCABIES, CRABS (PUBIC LICE), AND HEAD & BODY LICE

Many non-fatal but painful diseases are so common in the homosexual population that Shalit concludes "getting scabies, crabs, or head lice is part of being a sexually active" homosexual (1998, p. 232). An example is the common sexually transmitted parasite infestation called scabies (also known as *Sarcoptes scabiei*). Scabies is an "infestation," not an infection, because it is caused by an insect called *Sarcoptes scabiei*, not a bacterium or a virus (Chin, 2000). Doctors often diagnose the disease by skin examinations, via a skin biopsy, or by scraping. The burrowing bugs produce red or white swollen lines because they don't crawl on the skin, but burrow beneath the skin surface in 2 to 3 minutes after skin contact (Goldstone, 2001). Scabies burrow under the skin to make a nest to rear their young, whereas crabs (pubic lice) lay their eggs (nits) in the pubic hair. Most infestations contain ten mites or fewer, and the reservoir is humans. Other species of mites can live on humans, but do not reproduce on them (Chin, 2000).

A major symptom is intense itching that worsens at night. Scabies infects the genitals, breasts, arms, lower abdomen, and between fingers (the web spaces). The primary populations at risk are the sexually active, the immunologically compromised, and those living with or in close contact with infected persons (Dauden et al., 2000). Scabies is usually acquired from a partner during sex, but any form of close contact can enable the insects to transfer from one person to another (Dauden et al., 2000). One can sometimes catch scabies even by sharing clothing, towels, or a bed with an infected person, although it usually requires more intimate contact with an infected individual than a handshake (Goldstone, 2001).

The time period before symptoms appear is usually from 2 to 6 weeks for persons not exposed previously, and from 1 to 4 days for re-exposure (Chin, 2000). Itching starts from several days to up to 4 to 6 weeks after the infestation. For persons previously infested, itching begins sooner because the body has previously been sensitized (Goldstone, 2001).

People with HIV or other immune-compromising conditions are at risk for a very severe form of scabies called Norwegian or *crusted scabies*, which is highly infectious because of the large number of mites in the exfoliating scales (Chin, 2000). Although scabies itself is not lethal, it often indicates infection with another, potentially more serious STD. A major concern is that scabies can be complicated by hemolytic streptococcal infection (Chin, 2000).

Lice are larger than scabies mites and require humans for survival. They are very contagious—sex with an infected partner carries a 95 percent risk of infection (Shalit, 1998, p. 232). Many of the symptoms are the same as those for scabies.

THE PROMISCUITY PROBLEM

Not only does the type of sexual behavior in which homosexuals engage place them at a much higher risk of disease, but their high level of promiscuity also is a major contributor to their health problems (Cameron et al., 1989; Bolton, 1992; Cheong et al., 1997; Hegyi et al., 1997; Kafka and Hennen, 1999; Davidovich et al., 2000, 2001; Remis et al., 2000). One survey indicated that homosexual males have an average of *over 50* sexual partners in their lifetime (Rueda, 1982, pp. 52-53). Bell et al. (1981) found that 43% of White male homosexuals reported having sex with *more than 500* partners, and 28% with *over 1,000* partners in their lifetime.

A newer study found that 28% had over 1,000 partners, 15% had 500 to 1,000, 32% had 100 to 500, and only 25% had fewer than 100 partners in their lifetime (Cone, 1994). A major text on counseling noted that gay male relationships "lack the norms built into heterosexual marriage or other long-term male-female coupling, so that infidelity is not necessarily considered a serious offense or deviation."

The text concludes that "sex with a variety of partners is a fact of life in gay male relationships ..." and therefore we "should not apply heterosexual norms" to homosexuals (Goldenberg and Goldenberg, 1990, p. 194). They suggest that counselors should accept these different norms as "normal" and acceptable. Of course, this only condones behavior that often eventually proves lethal. Goldenberg and Goldenberg then conclude that a counselor who feels "uncomfortable" with this lifestyle should "work through any internalized underlying interfering homophobic feelings" (1990, p. 199). The authors stress that the role of a counselor is to help "gay" clients "accept, and integrate a gay identity."

The health risks of homosexuality are correlated not only with the number of partners, but are also a major problem even among those persons in long-term "monogamous" homosexual relationships (Cameron, 1998; Davidovich, et al., 2000; Meeker, 2002). Youth today are also adopting homosexual behavior sooner (often before age 14), exacerbating the problem. Meeker concludes that teens are today

> "coming out" at earlier ages. A few years ago, gay boys and girls tended to announce their sexuality between the ages of 18 and 22. Today, kids as young as 12 are declaring they're gay. I have two serious concerns about this trend. First, kids at 12 or 13 are still in the process of identity formation. How can they possibly know yet what their sexual orientation will be? Second, and more importantly, younger adolescents, no matter what their sexual orientation, are at high risk for having multiple partners and more diseases if they start having sex at an early age. Gay or straight, these kids are running a great risk. (2002, p. 153)

Cooper (2000) found that, of the groups he studied, homosexual men are also at the highest risk of becoming "cybersex compulsives," meaning they spend more than 11 hours per week on their computer for sexual purposes. The conclusions of surveys in this field depend upon the sampling population, sample size, and specific questions asked, but they all document that an enormous amount of promiscuity is a common part of the gay lifestyle (Cone, 1994). Also, the level of the disease problem can be debated, but there is no question that the problem is serious (with, until recently, AIDS being the most-publicized example).

This promiscuity clearly is contrary to the Biblical injunction that a man and woman marry and "not defile the marriage bed." Although promiscuity among heterosexuals also carries many dangers, they are generally far fewer than those associated with homosexuality. Infections caused by sexual relations are rare in monogamous couples who practice appropriate hygiene and normal sex. One reason is that vaginal secretions contain high levels of germicides that successfully minimize the chances of infection as a result of heterosexual sexual relations.

Conversely, similar secretions are not produced during either anal or oral intercourse. The major anatomical problems associated with homosexual sex (tearing of mucosa and lubrication problems, the latter often overcome by the use of various jellies) generally are not a problem in heterosexual relationships. One would expect equal protection for both homosexual and heterosexual acts if both were created by God.

Another problem is that many self-identified homosexuals engage in heterosexual sex, and, as a result, spread disease to the heterosexual population. One study of 498 self-identified lesbians found that fully 81% reported involvement in heterosexual behavior (Lemp et al., 1995). Furthermore, Masters and Johnson's scientific studies of persons labeled homosexual found that both *groups* consistently listed heterosexual encounters as highly erotic, actually at the top of a list of their erotic fantasies. In one study, both male and female homosexuals listed a "heterosexual encounter" as their *third* most common sexual fantasy! (McCutcheon, 1989).

This finding also supports the conclusion that most persons self-labeled as gay are, at best, in varying degrees bisexual—especially in view of the fact that many also have heterosexual relations and many were once married and had families. Psychiatrist Emmanuel Rosen concluded that "all people have both heterosexual and homosexual drives. What varies is how one deals with those drives. To have a homosexual impulse or fantasy has nothing to do with your sexual orientation (1998, p. 56). Many people become involved in a homosexual lifestyle after they are married and have children. Disease transmission among non-promiscuous heterosexual couples is rare and is almost always due to poor hygiene or infidelity.

CONCLUSIONS

One reason for many of the scriptural prohibitions (e.g., cleanliness, quarantine rules, etc.) was to help protect physical health. Likewise, the evidence indicated that a major reason why Scripture condemns homosexual behavior is also due to its detrimental health effects. The medical literature demonstrates that homosexual behavior has a clearly detrimental effect on health, causing a variety of serious and eventually lethal diseases. While this review focuses on males, persons of both sexes involved in lifelong homosexual practices are at risk, as shown by the fact that both male and female practicing homosexuals live only into their late-40s (Cameron, Cameron, and Playfair, 1998).

PART II:
CYCLOPEDIA OF NORMAL AND ABNORMAL SEXUAL BEHAVIOR

Introduction to Part II

Homosexuality is only one of over one hundred different sexual orientations. This cyclopedia covers the most common ones. This work evolved from our clinical work as licensed therapists at Arlington Psychological Associates and research psychologist employment for the Oakland County Court System. A common issue was the need to use correct terminology when dealing with sexual offences. As we encountered various forms of sexual behavior, we were required to classify them as part of our research and clinical obligations. We relied heavily on *The Diagnostic and Statistical Manual of Mental Disorders* (DSM) by The American Psychiatric Association.

The list and classifications do not make judgements about abnormality or political correctness but help to classify behavior by use of the correct terms. It is not exhaustive but rather classifies most of the common terms required for clinical or legal purposes. As this guide was prepared for professional work, pronunciation helps, especially in court. For this reason, we did not follow any standard guide. The *Society for the Scientific Study of Male Psychology and Physiology* was founded in 1979 by a group of college professors and clinical psychologists.

CHAPTER 9

Heterosexual Relations

TYPES OF HETEROSEXUAL BEHAVIOR

Coitus—Insertion and penetration of the female vagina by the male penis, also called **sexual intercourse**, **cohabitating**, **coition, copulation**, **to have sex**, **or to make love**. The specific act of a male inserting an erect penis into a female vagina is called **intromission**. Without birth control protection, it is called **bareback**. **Slang**: bang her up, bedtime story, be on the make, board her, bread a leg, carnal knowledge of, dip your wick, give her the business, get butt, chunk, crack, do, do her wrong, ease nature, fix her up, flesh session, flop her, frig her, frigging, fuck, get in, get fixed up, get some, get some ass, get some butt, get some jazz, give a frigging, hosing, give her a past, give her the business or works, go the limit or route, go to town with, hit it off, hop into bed with, hose, hosing, hump her, hunk her, jab her, jazzing, joy ride, jump her, knock her off or over, to know, to lay, lay the leg, let nature take its course, to make love, mash, make a killing, make Mary, make a dead set for, niggle, nookie, nooky, not do right by our Nell, party, to go up a drainpipe, peter (give), phutzing, get a piece (or hunk of tail), pick her cherry, plant my oats, play house with her, put the boots to her, ride her, roll her, roll in the hay, screw, screwing, scuttle, see the elephant, sexperience, sexpressions, shag, shoot one's wad, short-arm practice, short time, skirt, sock it to her, to swive, tail, take it out in trade, take on, take her virginity, tear off a piece or hunk of skirt, get the works, to 99.

GENERAL HETEROSEXUAL BEHAVIOR

1. **Kissing**—Any body contact where one person uses his/her lips.
2. **Smooching**—Sexual activity involving kissing, mostly for sexual arousal.

3. **Soul Kiss**—Mouth to mouth in which the tongue of either partner is inserted into the mouth of the other partner. Also called *French kiss*.

OTHER TYPES OF HETEROSEXUAL BEHAVIOR

1. **Anal intercourse**—Where the male penetrates with his penis the anus of either another male, or a female. Also called **anal copulation**, **anal-genital intercourse**, and **anal eroticism**. **Slang**: buggery, rear entry, Greek culture, have it Greek, bumming, ass or butt fucking, moon shot, one up the bum, stern job, bottling, back scuttling, corn holling (or balling), bending some ham.
2. **Auxiliary intercourse** (*aks zill e ary*)—Stimulation of the penis by the armpit of one's partner.
3. **Coitus a tergo** (*co i tus ah te air go*)—Vaginal intercourse with entry from behind the women.
4. **Coitus interruptus** (*co i tus in ter rup tus*)—Where the male withdraws the penis from the vagina before semen ejaculation, usually to prevent pregnancy, but also so he can ejaculate on the woman's body or elsewhere specifically for sexual reasons.
5. **Digital coitus**—Manipulation of one's partner's genitalia with one's hands, either homo or heterosexual. **Slang**: finger fuck, frigging, hand job.
6. **Interfemoral intercourse** (*inter' fah mor al*)—Coitus without actual physical penetration of the vagina, achieved by movement or rubbing of the penis between the female's legs. Although rare, pregnancy can result from this action. Also called **Inter-curual intercourse** and **femoral coitus**.
7. **Intermammary intercourse**—Where the male achieves orgasm by placing his penis in between the female breasts. A type of **pseudocoitus**. **Slang**: titty fuck, tit fuck.
8. **Oral intercourse**—Sex in which the mouth of a male or female is used to cause ejaculation in the male or orgasm in the female; see Fellatio and oral sex.
9. **Perineal intercourse** (*pair ah nee al*)—Where orgasm is achieved by the male rubbing his penis against female perineum.
10. **Pseudocoitus**—When the penis becomes stimulated and the person achieves orgasm by direct contact with the nude, or partly nude, female body but does not penetrate the vagina or other body orifice. This form of sex excludes anal, oral, vaginal, and interfemoral intercourse.
11. **Urethral coitus** is coitus where the penis penetrates the urethra instead of the vagina, rare except in cases of deformity.

12. **Vaginal intercourse**—Where the male penis enters the female vagina, usually or with the goal of achieving orgasm. The term coitus and sexual intercourse most often refers to vaginal intercourse only (See coitus.)
13. **Virginal**—One who has never had sex, applies to both males and females.

TYPES OR SOURCES OF SEXUAL STIMULI

1. **Altererotic** (*al ter air rot ik*)—any sexual behavior that involves one or more other persons.
2. **Autoerotic** (*au toe air rot* ik)—Sexual behavior involving only oneself and no other person.
3. **Endogenous** (*en dog en ous*)—that hetero or homosexual stimuli which originates within the person him or herself.
4. **Engram** (*en gram*)—when the source of sexual stimuli is a vivid memory of a past visual or psychic sexual experience or a prospective act. Also called **psychologny**.
5. **Erethism** (*ah re thizm*)—extreme sensitivity to stimulation to the point that light caress or soft touches are sufficient to cause vivid erotic desires or erotic responses. If normal touching causes nervous tension or irritability. This is also a form of hypertension.
6. **Foreplay**—Any physical stimuli designed to sexually arouse. **Slang**: work-up, breakfast in bed.
7. **Tactile** (*tak toll*)—any sexual stimuli that results from being touched, especially by the hands or lips, but also the feet or other parts of your or another's body. **Slang**: feeling, to cop a feel, feeling her up, roving hands, roving lips. That involving the entire body is called "going around the world."

TYPES OF MARRIAGES

1. **Concubine**—A woman who legally lives with a man who is not her husband, generally in exchange for her keep and protection. In some societies, wealthy or middle class married men may maintain one or more female household "servants" who are also openly available sexually. This arrangement is with the full knowledge and approval of the wife of the household. Only those women who are involved in such a relationship are concubines, but some use the term to describe any female who dwells with a man in exchange for her keep, or most of whose living expenses are paid by him.
2. **Kept man** (or woman)—Same as above, only not socially approved and done without the wife's (or husband's) knowledge or approval. **Slang:** sugar daddy, or sugar mommy.
3. **Digamy** (*die gah me*)—Refers to the social practice or act of entering into a second marriage after the first spouse dies.

4. **Endogamy** (*en dog ga my*)—Is the requirement that marriage be only to a person that is **inside** of some social group, such as a certain race, nationality, tribe, or extended family. Thus a person who cannot marry outside of his race is under an endogamy rule.
5. **Exogamy**—Refers to the requirement that marriage be only to those **outside** of some family relationship, such as outside of one's family group or clan. Violation of this social taboo is called **incest**, and an example is the prohibition against marrying one's cousins.
6. **Heterogamy** (*het ter og ah me*)—Marrying persons unlike oneself ethnically, socially or one of a different social class, such as the proverbial pauper marrying a prince.
7. **Homogamy** (*hah mog ah me*)—When persons must marry someone similar to themselves in terms of economic, social class rank or other traits.
8. Miscegenation (*mis ceh ga nah shion*)—An interracial marriage, in the West usually between a Black and White, a White and a Mexican or two other widely different races.
9. **Pantagamy** (*pan tah gah me*)—A rare practice now existing only in some small societies in which every adult male is regarded by custom or law as the husband of every adult woman and vice versa.
10. **Platonic marriage** or platonic love or a platonic relationship—A nonsexual marriage. The term "platonic," is derived from Plato, an ancient Greek philosopher who idealized emotive relationships uncharacterized by sexual desire and/or congress. It is commonly used to describe a nonsexual but emotional involvement under circumstances where sex would often be expected. Thus, in a "platonic relationship" the male and female parties are fond of each other but are not involved sexually. If the parties' fondness for each other is deemed especially intense, the relationship is described as involving "platonic love." The key element in platonic relationship is lack of direct sexual involvement, though it might ordinarily be expected, and one would not describe an emotive relationship with a parent, a child, a sibling, a twin, etc., or a relationship with a same gender friend as platonic.
11. **Polyandry** (*poly an dry*)—Refers only to a plurality of husbands, or a situation in which one wife is married to several husbands at the same time; in this system usually when a woman weds a man, she is automatically considered married to his brothers. This form of marriage is very rare and is usually found in very poor societies in which several men are necessary to support one woman. These societies also usually practice some form of female infanticide.

12. **Polygamy** (*pol lig ah me*)—Refers to a marriage plurality of either wives or husbands and not just wives as is often assumed. Includes polygyny, polyandry, and bigamy.
13. **Polygyny** (*po lig ah ne*)—Refers to a marriage plurality of wives or the situation where one man is permitted to marry several women.
14. **Serial Monogamy** (also called **serial polygamy**)—Is where a person can have only one wife or husband at a time; thus, to marry another, one must first divorce. This system is now the prominent form of marriage in the western world.
15. **Tetragamy** (*tet tra ga me*)—Is the term for a social system that allows males up to four wives. Mohammedan law purportedly permits four wives, but actually Mohammed had as many as fourteen wives at one time, and never specially limited his subjects to four. The four-limit idea represents an interpretation of his writings. Most Moslems, though, have only one wife.

CHAPTER 10

Levels of Acceptance of a Partner for Sex

Among the potential partners for sex includes some more socially acceptable than others. The following are some common examples, from least to most socially acceptable.

1. **Deviant**—Sex with anyone who is available at the time, regardless of their suitability or the local social norms. This category includes, among others, rapists, pedophiles, prostitutes, homosexuals who have sex in bathrooms, and stereotypic nymphomaniac behavior.
2. **Casual**—All forms of "recreational sex" involving very loose standards and, at best, limited use of force. Enthusiastic swingers are typical of this class. **Slang** (for females only): biddy, bimbo.
3. **Selective**—Requires some affection or feeling for the sex partner and tends to be more personalized than the above; the object is, importantly, a person, and made to feel such, and not a thing used solely for gratification.
4. **Affectionate**—Requires clearly defined, understood **emotional** attachment to the sex partner.
5. **Unique**—The "one and only" syndrome, limiting sex to one person only, one's steady girlfriend or boyfriend, or with a person whom one has had a long-term relationship and desires to continue it. If one lives with this person, it is called shacking up or living together. (See Liaison, chapter 11.)
6. **Marital Fidelity**—Sex with no one else but one's formal, legal marriage partner, called one's spouse. The Judeo-Christian-Moslem ideal.

CHAPTER 11

Illegal or Socially Disapproved Heterosexual Coitus

The section usually refers to illegal or socially disapproved heterosexual vaginal coitus. The most common example is **prostitution**, or sex that one pays money or services for, often with a person who makes a living by performing various sexual services for money, goods, services or one who supplements one's income by regular marketing of sex. They often specialize, or offer a wide variety of sexual services, such as bondage, etc. A male prostitute is a **Gigolo**. A high-class prostitute is a **call girl** or **call boy** or an **escort**. An older prostitute is a **Harridan**. A male who solicits homosexual partners in a public restroom is a t-room queen or a toilet queen. **Slang**: alley cat, ass peddler, baggage, bimbo blister, bucket broad, butt peddler, chippy, cash bag, cab moll, cash escort, Delilah, daughter of joy, doxy, fallen women, fancy quiff, floosey, flooze, grisette (French), human mattress, hotsy, hustler, gash, hooker, burlap sister, crack saleslady, hot piece, harlot, demimondaine, felle de joie (French), jade, a Jezebel, lady of the evening, nookie, pay for play girl, popsy, pro-lady, pross, prossy, minx, sin sister, Sadie Thompson, strumpet, sportin' woman, pickup for dough, trull, trucker's delight, pavement pounder, painted lady, puta (Spanish), putana (Italian), tart, tail peddler, working girl.

A home or place where prostitution regularly occurs is a **bordello**, or bordel. **Slang:** brawdy house, brothel, call house, cat house, flesh pot, fallen women house, house of ill fame, house of ill repute, house of pleasure, house of prostitution. A person who solicits customers is a brawdry, pimp, procuress, or hustler. A customer is a *trick, sap or John.*

OTHER TERMS RELATING TO SOCIALLY DISAPPROVED HETEROSEXUAL COITUS

1. **Adultery**—The act of coitus by a married person with someone other than their legal spouse. Also called **infidelity**, **unfaithfulness**, **extramarital sex,** and **extramarital relations**. **Slang**: screwing around, swinging, to cheat on, to have some on the side, to sly poke, to two-time, to run or play around, to get a slice, to yard, to horse around, to put the horns on her, to cuckold, to fool or mess around.
   a. **Adulterer**—A married male who commits adultery.
   b. **Adulteress**—A married female who commits adultery. Her husband is called a cuckold.
2. **Fornication** (*for nah ka' shun*)—The act of coitus by an unmarried person of either sex.
   a. **Fornicatrix** (*for ni ka triks*)—An unmarried female who often engages in coitus.
   b. **Fornicator** (*for ni ka* ter)—An unmarried male who often engages in coitus.
3. **Lecher** (*let sher*)—An adult who engages or tries to engage in sexual activity with females, especially those who are considerably younger than he is, and the male lacks tact and social graces or uses females.
4. **Liaison** (*li ah son*)—Where two people live together without being legally married. Also called **common law marriage**, **living together**, or **living together without the benefit of marriage**. **Slang**: shacking up, living in sin, rooming in, playing house.
5. **Libertinage** (*lib er tin age*)—Any sexual behavior which is inhibited by ethical or legal restraints. Other terms: libertinism, sexual license. The person is called a **libertine**.
6. **Licentious** (*lie cent shis ious*)—Speech, behavior, literature, etc., which is primarily and openly sexual in nature and would not be employed or practiced in "polite" company; thus "licentious" involves "vulgar" sexual vernacular; "licentious" activity is sexual behavior which would ordinarily be judged impermissible under contemporary community standards, and "licentious" literature is that which the primary focus of is sex. Also lewd, salacious, philanderer.
7. **Menage a Trois** (man wah ah Trog)—A married couple who add a third person into their sexual relationship, producing a "three-some." This person can be of either sex, but is usually heterosexual, or only male or female homosexuality behavior predominates. **Slang**: house of three, threesies.

a. **Affairs**—a form of *Menage a Trois* only done clandestinely, without the spouse's knowledge. **Slang**: cheater, two-timer.
   i. **Cicisbeism** (*ce ces ba ism*)—Refers to a female who has an affair or a regular sexual relationship with a male other than her husband, often of the same social class.

9. **Paramour**—A long-term non-spouse sexual partner involved in a heterosexual relationship with a married or single person.
10. **Promiscuous**—Meaning has few long-term relationships or is married and has many short-term relationships, often primarily for sexual gratification. **Slang**: whoring around, sow one's wild oats.
11. **Whoremonger**—A male who seeks out very unappealing females for sex.

STIMULI FROM OTHER PERSONS; TYPES

1. **Ephebophilia** (*e fe bow fill e ah*)—an erotic drive or sexual desire by an adult for adolescents. Very common, and often not seen as abnormal, although it is often illegal with a preadolescent, behavior which is then called **pedophilia**.
2. **Gerontophilia** (*jer on to fil e ah*) (*Greek, geron* = old, old man; *philia* = love)—when the choice or preference of a young person for sexual objects or partners who are elderly persons of the opposite sex. Usually the age difference is 40, 50, or more years, and the younger person may be under 25 (usually a male) and the older person (usually a female) is 70 or older. The younger person involved in gerontophilia is a **gerontophiliac**. This is a rare sexual disorder, less than 0.01% of the population, and the most common reaction is not a feeling of revulsion or shock, but surprise.
3. **Incest** (*in cest*) (*Latin, incestum* = unchastity)—Sexual relations with any other family member except one's husband or wife, usually between the father and his daughter or step daughter but includes a male with his sister and a mother with her son. The word incest usually refers to **heterosexual** relations but can include homosexual as well.
4. **Incubus** (*in ku' bus*)—The act of a fallen angel, evil spirit, or a demon indulging in sex with persons in their sleep.
5. **Pederasty** (*ped er as ty*) (*Greek, pedo* = child)—Insertion of an adult penis in the anus of a child or a legal minor for the purpose of sexual intercourse; sodomy by adults on children. A person who commits pederasty is a **pederast. Slang**: honeydipping.
6. **Pedophilia** (*ped oh feel e ah*) (*Greek, pedo* = child; *philia* = love)—Real or imagined sexual intercourse or activity by an adult with a young child, usually of the opposite sex. There is usually a great age difference, such as a ten-year-old having coitus with a 45-year-old. If with an adolescent, it is

called **Ephebophilia** (*e fe bow fil e ah*). A person who commits pedophilia is a **pedophiliac,** also spelled **paedophilia**. Also called **infantosexuality**, and a person who indulges is **infantosexualist**. **Slang**: honeydipping, cherry picker, chicken farmer, cradle snatcher or robber, having jail bait.

7. **Prostitutephila** (*pros tah too tah fil e ah*)—The desire for sex specifically with a person who sells sexual favors. The term usually refers to a male client and a female serving.
8. **Sodomy** (*sod ah my*)—from the Bible city of Sodom, destroyed by God because of its sexual sinfulness (Genesis 18–19). The act of anal copulation or insertion of the penis inside the anal opening of either a male or female. See anal intercourse. A person who commits sodomy is a **sodomist**.
9. **Spectrophilia** (*spect tro fil e ah*)—The act of having sex with demons or wicked spirits. A person who desires such is a **spectrophiliac**.

GROUP SEX, TYPES

1. **Daisy chain**—When the persons involved in group sex are physically "linked" in sex, performing on others while being performed on.
2. **Group Sex**—A form of sexual behavior in which three to four people involve themselves in specifically sexual activities at the same time and in view of each other; thus observation, or elements of voyeurism are often involved, but only indirectly. **Slang**: groupies.
3. **Orgies**—A form of sexual behavior in which a group of persons, usually five to twelve or so, participate in sexual activities, both at the same time and in view of each other. Observation is often a part, but not the total sexual stimuli or activity of most persons involved. **Slang**: Roman games.
4. **Pluralism**—A form of sexual behavior in which a large group of people participate in social behavior which includes various "sexual fantasies." The number of participants may range from a dozen to as many as several hundred or more. Participants tend to display both a total lack of inhibitions and a disregard for conventional methods of sexual gratification, i.e., they often participate in group homosexual sex, oral sex, etc. Sexual pluralism is also known as **sexual bohemianism**, or **sexual anarchism**. Sexual pluralism usually involves more persons than an orgy (orgies include around five to a dozen, group sex three to four persons).
5. **Communal Sex**—In a group living situation, when those involved can share (actually trade) sexual partners or, if some permanent pairing arrangement exists, can swap wives or partners.
6. **Train Sex**—A woman or man who has sex with several persons in a row. A vernacular term for the female who engages in coitus with one man

immediately or shortly after she has performed coitus with another is a **buttered bun** or a wet deck. The former term comes from the fact that her genitalia, slick with the first man's semen, is described metaphorically as being a "buttered bun." **Slang**: "do a train on," gang bang.

CHAPTER 12

Exhibitionism

Exhibitionism (*eks ah bish an ism*) (*Latin, exhibere* = to exhibit)—Self exposure, usually specifically of one's genitalia with an erection to a stranger for one's own sexual excitement. Most arrests involve males, although when women express this behavior, they are far less likely to be arrested. The source of the erotic response for males, especially when strangers are involved and especially young girls or females in general, often is primarily the victim's response (shock, fear, outcry, etc.). Ignoring or reacting in an unshocked manner, telling the person to "put that thing away," often helps to reduce or stop, for that time at least, the exhibitionists erotic response. Sexual exhibitionism by a female is called a **beaver shot**, **beaver show**, or **to show pink** and is considered more socially acceptable by both sexes. This behavior is a form of the **spectacular complex**. The following are the basic types or forms of exhibitionism.

1. **Adamism** (*ah dam izm*) (Hebrew, for *Adam* or **man**; the first man created, thus completely nude)—A form of exhibitionism where the subject exhibits him or herself completely nude, and not just his or her sexual organs as is usual in exhibitionism.
2. **Exposure**—When the person exhibits only *part* of his body, usually the genital area, by taking out his penis, for example, or showing his buttocks. The classic form is a man opening his trench coat to show, primarily, his erection.
3. **Inspectionism** (*in spec shion izm*)—Sexual pleasure from looking, specifically at the genitals of another individual.
4. **Pornography**—A class of pictures, magazines, books, literary works, sculpture, painting, films, which the main focus is on sexual behavior of all types, which titillates or sexually arouses the reader/viewer and which often treats sex in a manner that is often deemed inartistic or even offensive, and the only purpose is sexual arousal. The word "pornography," *Greek*, *porneia*,

meaning "brothel" and *grafo*, meaning "to write") is, etymologically speaking, literature pertaining to prostitution, or to the lives, relationships, and sexual activities of prostitutes and their customers. The term as used today describes a wide variety of artworks that depicts sexual relationships and activities. In the early twentieth century, many described as "pornography" any artwork that contained vivid sexual detail or which appeared to be primarily sexual in focus. Subsequently, the term has been narrowed to include only those works of which the primary purpose is to titillate or sexually arouse the reader/viewer, and/or which treat sex in a manner deemed inartistic and offensive. A pornographic film is called a stag film.

a. **Lesbian** "romance" is called clit lit. **Slang**: *Smut.*

5. **Scoptophilia** (*skop to fil e ah*) (*Greek, skopes* a spy or watcher; *philia* = love)—Similar to voyeurism, except that it is the primary (or a very important) source of sexual stimuli for the person; often totally replacing heterosexual relations. Often this form involves looking specifically at sexual organs or sexual activity. A person who is addicted to scoptophilia is a **scoptophiliac**.
6. **Troilism** (*tr oi al izm*) (*French,* onlooker, spectator)—A form of exhibitionism in which a male desires to have sex with two or more partners at one time, or in the presence of several females. The common pattern is for two people to have sex while a third person watches, often a husband arranges for his wife to have sex with another person so he can watch.
7. **Voyeurism** (*voi yer izm*)—The act of looking at girls or women whom one does not know, and who are in some stage of undress, usually without them knowing; both the secrecy and illegality of the act is important. Often the potential of getting caught is an important part of the syndrome. A male who commits voyeurism is a **voyeurist** and a female voyeur is a **voyeuse**. A person who has a very strong drive to be involved in this is a **Voyeur Compulsive**. Observing one's lover in various stages of undress is not voyeurism because the above elements of secrecy, etc. must be present. **Slang**: peep freaks, peeping Tom, a looker.

CHAPTER 13

Masturbation

Masturbation (*mas ter bay shion*)—Where the person achieves orgasm, or tries to, without a partner by self-stimulation of some sort. In extreme forms, it involves both self-voyeurism and narcissism (narcissism is excessive love of oneself). The subject may use his or her own body image to become sexually stimulated, usually via a mirror. The most common source of stimuli for males is nude or erotic photographs of the opposite sex. The term more commonly refers to self-stimulation, that which culminates in orgasm also called **ipsation** or **ipsism**; older technical terms include **Urethrorrhea ex libidine sexuale** or **coitus interdigitae**. All older forms of self-stimulation are collectively termed autoeroticism. A person who masturbates is a **masturbator**. **Slang**: hand job, jack off, jerk off, J/O, to beat one's meat, to squeeze off, to whack off, to wank, to toss off, to choke the chipmunk, to play solitaire, to pound the pud, to milk, to pull the pudding, to punish Perry in the palm, to have one off the wrist, to touch, to oneself-up, to jerk the gherkin.

Old terms, now slang, are the *solitary sin*, and *self-abuse*. Another common antiquated term is **onanism** (*on nin ism*) from the Bible character *Onan* (the word in Hebrew means "vigorous") found in Gen. 38: 6–10, 46: 12. When his brother Er died, Onan was to impregnate Er's childless wife in order to carry on the family name, as per Jewish law, but "wasted his semen on the ground rather than to produce an offspring," evidently so as he would receive Er's inheritance instead of Er's wife's child, if she became pregnant. Actually, what was involved was not masturbation, but *coitus interruptus*, and the sin was a violation of Hebrew family, not sexual, laws.

MAJOR FORMS OF MASTURBATION

1. **Autocunnilingus** (*ah to kun nil lin gus*)—Masturbation by a female stimulating her own vulva with her mouth, hips and tongue, a difficult feat that some women practice for years to achieve.
2. **Auto-fellatio** (*ah to fel lay she oh*)—Masturbation where the male stimulates his penis with his own mouth, tongue, and lips., a difficult feat which some males practice for years in order to achieve. Also called **auto** or **self-irrumation**.
3. **Autoerotic asphyxia** (*au to er otic as fix ee ah*)—Masturbation which is accompanied by a reduction in one's usable air supply (often by partial hanging) so as to increase sexual pleasure. Oxygen deprivation, some claim, can heighten sexual pleasure but is dangerous, especially if done alone. A set of ropes is set up, almost as if one was planning to die by hanging, except that the person intends to stop before death. If something goes wrong, such as the person blacks out too soon, death may result. For this reason, all "suicide by hanging" must be fully investigated ruling out this possibility.
4. **Physical masturbation**—Sexual excitement obtained primarily by mechanical stimulation of the genitals, in contrast to **psychic masturbation**.
5. **Psychic masturbation**—Sexual excitement obtained primarily, or mostly without mechanical stimulation of the genitals or other body parts. Usually visual or pictorial stimulation is used (thinking of erotic thoughts, viewing pornographic pictures, reading pornographic books, or viewing erotic movies or realia) although some body movement may be present.
6. **Psycho-sexual infantilism**—Where an adult remains at a child's sexual developmental level. Sex for persons in this category involves mostly physical manipulation of the sexual organs with little thought of one's partner, if one has one.
7. **Unconscious ejaculation**—is the expulsion of sperm from the penis by muscle contraction caused by a variety of both sexual, non-sexual, conscious and unconscious factors. The medical term is **spermatorrhea** (*sper mat or re ah*). Many internal and external factors, both sexual and nonsexual, have been identified which can cause male erection and/or ejaculation. This term is concerned with any internal or external physiological factors which are influential in causing nondeliberate autoeroticism and unconscious ejaculation. **Slang** terms for any form of penis ejaculation: shooting jism or one's wad, giving milk showers. Slang for sperm includes: seed, jism, milk, semen, and spunk.

TYPES OF UNCONSCIOUS EJACULATION

1. **Nocturnal emissions** are unconscious erotic dreams or thoughts which cause an erection and, not uncommonly, orgasm, without physical manipulation of the sex organs. If caused by daytime thoughts, it is called *diurnal emissions* and is described above under **psychic masturbation**. Only those that occur while one is sleeping are formally nocturnal emissions. Another, older term is **oneirogmus**. **Slang** terms for nocturnal emissions include wet or erotic dreams, midnight glory.
2. **Defecation spermatorrhea** (*sper mat or re ah*) are ejaculations caused by, or influenced by, defecation. If it is conscious, often an associate or other party is present, but this person or persons may not consciously be part of the activity. During adolescence, a major opportunity for masturbation is while defecating, thus this fixation sometimes develops.
3. **Postatorrhea** (*post tah tor re ah*)—A seminal discharge which results purely from the pressure that the prostate gland exerts on the other seminary structures. It is often primarily medical, due to inflammation, pressure, cancer, or tumors, etc. An erection may or may not be present when it occurs.
4. **Urination spermatorrhea** (*sper mat or re ah*)—Where erection and orgasm is influenced or caused by or associated with the act of one urinating Those factors that are typically present in defecation spermatorrhea are often present in this condition.

CHAPTER 14

Sexual Drive Levels

Sexual Drive Levels (see also Sexual dysfunctions) What is considered a "normal" sexual drive level varies in western society, but usually normative data is the bench mark. Any high sexual drive is **hypersexuality,** the opposite is **hyposexuality**. In the late nineteenth century, many sexologists believed that a certain sexual drive level existed in females which was "normal," and females who had possessed a higher level suffered some sort of psychopathology. "Andromania" was the condition in which a female appeared to possess a higher level of sexual desire than they themselves believed normal, and these females were labeled "andromaniacal," even some who merely desired coitus and/or other sex acts more frequently than once every twenty-four hours, or more frequently than a given male partner was desirous, or who desired a variety of sexual partners over a relatively brief span of time, or who could not achieve orgasm no matter how frequently they engaged in sexual congress.

Sexual desire in females, as well as in males, is a phenomenon which is subject to a great variety of physical and psychological influences, and the intensity of a given male's or female's sexual appetite is, of itself, rarely psychopathological. Because of the vague criteria as to what constitutes "extreme," "excessive," or "insatiable," many researchers try to avoid the use of the term "andromania" or its more popular equivalent "nymphomania" except as a negative or positive connotation. Others use the term highly subjectively; hence, a female described as suffering from andromania will, in the opinion of another, be evaluated as "normally sexed." The reader who encounters the term "andromania" must examine the context in which it appears to discern the writer's criteria for use of terms such as extreme, or excessive. Also called **hyperaesthesia**, **Lagnosis**, **insatiable lagness for females**.

SEXUAL DRIVE LEVELS

1. **Anhedonia** (an he don e ah)—The complete or almost complete lack of pleasurable feelings in the areas or organs which normally produce an erotic or pleasurable response. A person who has no sexual feeling in the nipples, skin, lips, or sexual organs is suffering from anhedonia. Lack of sexual feeling in the penis or clitoria alone is called sexual anesthesia.
2. **Chronophilia** where the person's sexual drive is not equal to their sexuoerotic age but is concordant with the age of his or her partner. An example is gerontophilia, sexual attraction of a young man to the elderly.
3. **Concupiscence**—Any sensual or sexual desire in general. **Slang**: horny, hot, hot nuts, hot pants, hot to trot.
4. **Continence**—A voluntary abstention from sex for social, religious, personal or other unneurotic, rational reasons.
5. **Erotomania** (air oh to main e ah)—A male or female with an abnormally high or insatiable sex drive. Also called **tentiginous** or tentigo. A person so affected is an erotomaniac. In females the condition is called nymphomania, in males gynecomania, discussed below.
6. **Frigidity** (*fri gid ah ty*) (*Latin, frigidus* = cold)—Inability or difficulty in engaging in sexual intercourse, usually because of emotional factors. The term often applies only to females. In men, the condition is called **impotency**.
7. **Gynecomania** (*gi ne ko main e ah*)—An excessive and uncontrollable sexual drive in males; **male hypersexuality**. An older term is **satyriasis** (*sat ah ri ah sis*) (From Greek mythology) a man of an extremely lustful nature. **Slang**: a Don Juan, Cassanova, wolf, animal, dirty old man.
8. **Impotence** (*im pah tence*) (*Middle English*, weakness)—Inability to engage in sexual intercourse, usually because of emotional factors. The term usually applies to males. A person with this problem is **impotent**. **Slang**: limp, no money in the purse, droopy, at half mast, with brewer's droop.
9. **Libido** (*lah be do*)—The sex drive. Adj.—libidinal.
10. **Nymphomania** (*nimf oh may ne ah*)—A morbid or high level of sexual desire or drive in the female; **female hypersexuality** or **andromania**. A female with this condition is a **nymphomaniac**. **Slang**: a messalina. See also Erotomania.

CHAPTER 15

Sexual Dysfunctions

Before describing sexual dysfunctions, a discussion of potency and impotency is required. This is by far the most common sexual concern.

Potency is a condition characterized by ability of a male to achieve penile erection, sustain penile erection to orgasm, and withhold orgasm until a coital partner has achieved her orgasm. With the onset of puberty, males acquire the ability to achieve penile erection from specific erotic stimuli and to ejaculate semen. They normally retain this ability until death, except when physiological disorders or psychological factors interfere. The ability to achieve penile erection and to ejaculate semen is referred to as "potency" and the inability as "impotency." The term, "impotence" is also used to describe a variety of impediments that interfere with achieving penile erection and orgasm.

This inability may be the result of physiological or psychological factors, such as guilt, anxiety, or a particular set of circumstances. For example, a male might find himself incapable of achieving penile erection in an automobile parked on a busy street but perfectly capable in the privacy of his own bedroom. A male who cannot achieve penile erection under any circumstances, for physiological reasons, and whose condition is diagnosed as irreversible is said to suffer "*permanent* impotence," and one who should normally eventually regain the ability to achieve penile erection suffers from "*temporary* impotence."

The problem of inability to achieve penile erection because of anxiety about a particular situation, fear of failure, or because of lack of sexual desire for a particular partner is labeled suffering from "situational impotence." "Potency" refers specifically and exclusively to the ability to achieve penile erection and to ejaculate semen, whereas "virility' refers specifically to the ability to sire children and is thus synonymous with "fertility." A male may be potent without being virile and may be virile (meaning produces fertile sperm cells) without being potent (because he can sire children by artificial insemination).

COMMON SEXUAL DYSFUNCTIONS

See also Sexual Drive Levels, chapter 14 and Sexual Phobias, chapter 23.

1. **Anisomastia** (*ah nis oh mast e ah*)—Unequal size of female breasts caused by disease or use as for breast feeding.
2. Anorchism (*an or kiz m*)—A congenital abnormality in males causing the total absence of testicles.
3. **Anorgasmia** (*an or gas me ah*)—Is the inability to achieve orgasm under any conditions. Women able to come to orgasm only during solo masturbation, or who require much clitoral stimulation or oral contact during intercourse, are **semi-anorgasmic**. A wide variety of female sexual levels and response exists, and one would not be anorgasmic unless one was unable to come to orgasm under any conditions. In the male, anorgasmia is the inability to achieve any orgasm, including delayed ejaculation. Also called **anesthesia sexualis**.
4. **Aspermatism**—Male sterility from an inability to ejaculate or produce sperm cells.
5. **Atelia**—Incomplete genitalia development resulting in the sex organs looking like a child's.
6. **Clitoral impotence**—The inability to sexually respond in females.
7. **Coital aninsertia**—For males, the inability to insert the penis into the vagina of their partner; for females, the inability to be penetrated by a penis.
8. **Dyspareunia**—Any difficult or painful coitus, regardless of the origin or cause. Often it stems from an infection or a related medical condition or a psychological condition.
9. **Erectile impotence**—The inability for a male to have an erection.
10. **Erotic apathy**—A total or almost total lack of interest in any form of sex and a long-term lack of sexual desire, usually for psychological or social reasons, or lack of an attractive partner. This is one of the most common complaints of both women and men.
11. **Eviration**—Loss of masculine physical traits due to aging and the decline of sex hormone production. It usually starts in one's 60s and is complete in one's 80s.
12. **Gynatresia**—A physical closure of part of the vagina due to abnormality or disease.
13. **Mentulagra** (*men too lag rah*)—Condition in which a male experiences prolonged or perpetual penile erection, usually unrelated to sex. It can be very painful and requires surgery or medicine.

14. **Monorchid** (*mon or chid*)—A male who is born with only one testicle.
15. **Penile anesthesia**—A lack of sexual or any feeling in the male sex organ.
16. **Priapism** (*pri ah pizm*)—A physical condition where a constant erection of the penis occurs, usually the result of physiological factors and unrelated to sexual desire. In many cases the victim is incapable of normal orgasm. A male who experiences a constant erection is a **priapist**. Priapus—In Roman mythology, the god of "natural reproduction," son of Venus and Bacchus, possessed constant penile erection.
17. **Vaginal dryness**—A persistent failure of the vagina to lubricate or to maintain its lubrication. Some women solve this problem with the use of lubrication aids such as KY jelly. Others learn to assert themselves and demand longer foreplay.
18. **Vaginismus** (*vah gin is mus*)—A condition in which the muscles of the female sex organ contract too quickly. This causes the premature onset of what sexologists call the "postorgasmic refractory period" and effectively excludes the penis.
19. **Vulval anesthesia** (*vul val*)—A total lack of sexual feeling, or sometimes any feeling, in the female sex organ.

CHAPTER 16

Specific Noncoitus Sexual Acts

Sexual behavior that is judged indicative of emotional or mental disorders in the practitioner is termed sexual **paraphilia**. The term, "sexual paraphilia," gained wide usage in the late nineteenth century when many sexologists believed that indulgence in any orgasm-producing sex acts other than coitus—and a great many other, nonorgasm-producing sex acts—indicated, or was itself, an emotional or mental disorder.

In the scientific literature of the period, one finds statistically common sexual practices such as fellation, cunnilingus, and even masturbation, all described as sexually paraphiliac, along with the less common practices such as flagellation, transvestism, voyeurism, and exhibitionism. Among contemporary researchers, considerable disagreement still exists as to which sexual practices are in themselves indicative of an underlying emotion or mental disorder. General professional opinion is that most (but not all) variations are not psychopathological, especially if isolated and nondramatic. Only when the behavior is both problematic and a repetitive pattern, or the only mode of expression or satisfaction, does it constitute a clear psychosexual disorder.

TYPES OF NONCOITUS SEXUAL ACTS

1. **Anallingus** (*ay nal ling us*) (also spelled **anilingus**)—A positioning of the mouth to the anus for sexual reasons. Also called **analinctus**. Use of the tongue in this way is called **analinguism**. **Slang**; ass-licking, cleaning up the kitchen, rim shot on the job, tongue sandwich, rimming or reaming. A person who does this is an **aniliguist**.

2. **Oralism** (*or al* izm)—Any oral sex, or that involving the mouth on the sexual organs of the partner, includes both fellatio and cunnilingus. **Slang**: French culture.
3. **Lecheur** (*lesh ur*)—One whose total or primary sexual outlet is oral sex to the degree that other sex interests are either not developed or are sublimated.
4. **Cunnilingus** (*kun ah ling* gus) (*Latin, cuneus* = wedge, tongue) Positioning of the tongue or mouth to the vulva, usually the male to the female, for sexual reasons. From the woman's standpoint, the act is completed when she achieves orgasm.

   Other terms: **penilinctus**, **penilingus**, **penisuction**, **oral copulation**, **cunnilinctus**, and the outdated term **buccal onanism**. **Slang**: to French, Frenching, French culture, to lick pussy, cunt sucking, cunt lapper, eat her, go down, lap it up, muff diving, box lunch, having or eating cake box lunch, clam diving, eating at the Y. The slang term for a male who enjoys cunnilingus is called a fish queen.
5. **Fellatio** (*fel ay she oh*) (*Latin, fellare* = to suck)—Oral copulation or the female using her mouth and tongue to stimulate the man's penis, its use as a substitute sexual organ. *Latin, fellare*, mouth-penis. Also called **irrumation**, and the outdated term **buccal onanism**. The female who performs this act is a **fellatrice**—a male is a **fellator**. **Slang**: blow job, suck off, going down, to French, flute, head job, to give or get head, BJ, deep throat, sucking off, plating, basket lunch, copping a joint, Derby picnic, a Hoover, Eating out.
6. **Feffaire** (*fel lair*)—A female who assumes the oral or receptive role in fellatio. Also called **fellatrice** (*fell ay tres*).
7. **Oral genitalism** (*or al gen it till izm*)—A general term that specifically refers to all oral techniques of genital excitation or stimulation.
8. **Paresthesia, sexual**—Any type judged desire or appetite which is an indication of emotional or mental disorder or which is considered statistically atypical. Also called **aberrant sex**.
9. **Soixante-Neuf** (*Soi ante Nuckf*)—Where both partners perform oral sex on each other at the same time. Usually heterosexual, but the term also applies to homosexual encounters. **Slang**: 69 (*Soixante-Neuf* is French for 69), double wedding.

CHAPTER 17

Sexual Conversions

TYPES OF SEXUAL CONVERSIONS

1. **Ambisexual** (*am bi sex u al*)—A stage in one's very early development during which the person is **neither** male nor female, but has the capacity to become either, depending on the presence of the factor that causes sexual development, the TDF gene (its presence causes a male to develop, it is most always on the Y chromosome); a fetus at this stage is called an **ambisexual**. If a person is sexually developed but is lethargic, or emotionally uncommitted to a specific sex, he or she is said to be **bi** or **ambisexual** or sexually ambivalent.
2. **Castrophilia**—Where the male feels so strongly that he should be a female, that a desire to be castrated results, also called **Eonism** and psychic **hermephroditism**. A person with these drives, if strong, and persistent is called a **castrophiliac**.
3. **Bisexuality** (*bi sex you al* ity)—The desire to engage in both homo and heterosexual sexual activities. A person who is erotically attracted to both males and females is a **bisexual**. Also called **amphigenic** (*am phi gen ic*). **Slang**: AC/DC, Bi, a two-way person, ambi, switch-hitter, double gaited, and bi-minded. One who enjoys or practices performing oral sex on both sexes is called **bi-lingual**.
4. **Eunuch** (*you nick*)—A castrated male, a past practice with young boys in Medieval Europe so that their voice would not deepen, so that they would remain sopranos. Slaves and others were sometimes made eunuchs so as not to be able to have sex with local free women or others. A eunuch is usually heterosexual and sometimes will marry. The reason for becoming a eunuch

today is for sexual purposes, such when the castrophilia complex is carried out successfully.

5. **Gynandry** (ji nan dre)—The extreme grade of degenerative homosexuality in females.
6. **Hermaphrodite** (*her maf ro dit*) (*Greek, Hermaphrodites*—In Greek mythology, the son of Hermes and Aphrodite, when bathing, became united in a single body with the female nymph Salmacis, thus became both male and female). This condition is biological abnormality, either where the patient has part or the entire external genitals of both sexes or has male masculine genitals, ovaries, and female breasts. This person produces large amounts of both male and female hormones. A person who fits this category and responds erotically to both sexes is an **epicene**. Also called **gyandromorphism** (*gi an dro morphism*).
   a. **Pseudo-hermaphrodite**—A condition where the person has the **genitals** of one sex and the gonads (gamete-producing organs, which make sperm for males and eggs for females) of the opposite sex.
   b. **Androgyne** (*an dro guy ne*)—A female who possess certain physical male sex characteristics, due to hormone or other physical problems. Also called **androgynous gyander**. Or which a person of one gender possesses certain sex characteristics of the other, a female who possesses an inordinately large clitoris which resembles a penis, or a male whose testicles are undescended and whose penis is so small that it appears to be a clitoris. Genuine hermaphroditism is the condition in which an individual possesses the sexual characteristics of both genders to such a degree that gender determination is not possible on the basis of visual and tactile inspection alone; also synonymous with **androgynism**. Other terms: **Gynandrism**, or **gynandra invert**.
   c. **Androgynoid** (*an dro guy noid*)—A male who possesses one or more physical traits of a female, such as large breasts, hairless skin, or small sex organs. Also called a **gynandroid**.
7. **Lesbian** (*lez bi an*) (*Greek, Lesbios*, an island famous for female homosexuals)—A female homosexual. **Slang**: Queen, Nance, Viragos, Sapphism, a Sapphist, Dyke, Bull or Butch Dyke, Fish, invert diesel, radish, femme, Amy John, bull dagger, or bitch.
8. **Metatropism** (*met tah tro phism*)—Heterosexual persons with marked traits of the opposite sex—virile women, or weak, feminine men. A person with these traits is called a **metatrophiac**.

9. **Metamorphosis sexualis paranocia** (*meta mor fus seks yoo ah les par ah na oh see ah*)—A paranoid delusion or irrational fear that one's sexuality or sex has been physically changed, or soon will be, into that of the opposite sex.
10. **Transeroticism** (*trans ah rot ah cizm*)—This condition is either a fetish for, or a strong desire to, assume the name and role of the opposite sex or that which is the opposite of one's biological sex in a homosexual relationship. Usually, the male assumes the female role or dress. **Slang**: going in drag, a drag queen. A drag queen is often one who impersonates a female and performs in a night club or other entertainment place. A drag show is where several drag queens perform.
11. **Transsexual**—A strong belief that one is a member of the opposite sex, or would like to become the opposite of one's biological sex, usually males who want to become females. Sometimes these persons undergo "sex change" operations, involving hormone treatments, castration and surgical creation of an artificial "vagina." The person often wants to become a normal person of the opposite sex, and it involves much more than cross-dressing, labeled an extreme transvestite. **Slang**: TS, Swedish number.
12. **Transvestite** (*trans vest ite*)—A person who finds erotic satisfaction wearing and often displaying in public a total set of clothes of the opposite sex. Many put on makeup and shave their legs and underarms in order to look as much like a woman as possible. A transvestite is usually a homosexual who enjoys cross-dressing only and does not actually want to be the opposite sex. **Vernacular**: female impersonator. **Slang**: T.V., a drag or paint queen, a flip.
    a. Homeovestism. Sexual arousal from persons wearing clothing appropriate to one's gender, the opposite of becoming aroused by wearing clothing appropriate to the opposite gender.

CHAPTER 18

Homosexuality

Homosexuality is preference for sexual relationships or erotic attachments with someone of the same sex. Terms all meaning "homo-sexual" include **sexual invert, homogenic love, contra-sexuality, homo-eroticism, gay,** and **the third sex**. Older terms include: **simil-sexualism, uranism, sexual inversion, intersexuality transsexuality, psychosexual hermaphroditism** and **amphigenic invert**. An exclusive homosexual (Kensey scale, rating number 1) is a **paederast,** the lifestyle; **peaderasty**. Specific **slang**: **Dinge queen** is a black homosexual or male who engages in homosexual behavior. **Sea food** is a Navy homosexual, **dogs lunch** is an unattractive homosexual, **antie** is a middle-aged man seeking homosexual contact (this life is much easier for young homosexuals; older homosexuals often have difficulty finding partners.) **Slang** terms for male homosexuals include queer, fag, faggot, fairy, flit, fruit, daisy, Nellie, fem, belle, swich, radish, pouve, bent, woofter, pansy, Nancy, Molly, painted Willie, swish. For slang terms for female homosexuals, see **lesbianism**. Heterosexuality is sometimes called AC/DC; lesbianism, AC/AC; male homosexuality, DC/DC.

TYPES

1. **Amor lesbicus** (*Latin,* lesbian love)—Heterosexual male that enjoys watching lesbians make love.
2. **Bronco**—An adolescent boy who is resisting initiation into the homosexual life.
3. **Butch**—A female homosexual who takes the role of a male in a homosexual relationship and often in both coitus and social relations. The opposite of **Femme**.

4. **Dionism** (*di on ixm*) (*Greek, Dione*, mother of Venus)—Heterosexuality proper, especially in contrast to homosexuality.
5. **Femme**—A female homosexual who takes the role of a female in a homosexual relationship, both in coitus and in social relations. The opposite of **Butch**.
6. **Group Homosex**—The participation of several couples in simultaneous coitus and in view of each other. Involves elements of both voyeurism and exhibitionism, heterosexuality, and homosexuality. **Slang**: love-in, gang bang, group grope, daisy chain, cluster fuck, circus, gang shay.
7. **Tribad** (*trib id*)—A woman with an abnormally large clitoris and who plays the part of a male in homosexual acts.
8. **Triolist** (*tri oll ist*; tri = 3)—A triangular sexual relationship where one's opposite sex lover is a homosexual. Usually, the triolist watches his female partner (his wife or girl friend) making love with a female homosexual partner.
9. **Uranism** (*ur an izm*)—An obsolete term for the condition of homosexuality.
10. **Urnings** (*oor ning*)—An obsolete term for male homosexuality.
11. **Urninde** (*oor nin de*)—An obsolete term for female homosexuality.
12. **Viraginity** (*ver ah gin ah ty*) (from virago the older term)—Where a female has both the sexual feelings and mind of, or plays the role of, a male.

HOMOSEXUAL BEHAVIOR

1. **Buggery**—Anal intercourse by a male homosexual. A person who does this is a **bugger**. **Slang**: Belly queen, and the receiver is called a bender.
2. **Effeminate**—Pertaining to homosexual behavior, mannerisms or practices, or to individuals who engage in homosexual activity. **Slang**: lavender.
3. **Oncer** (*on cer*)—Where a homosexual desires no more than a single sexual encounter with each partner, a common trait of many male homosexuals, partly because this behavior does not obligate them to develop commitments.
4. **Tribady** (*trib ah* dy or *trib ah dixm*)—A homosexual act involving mutual friction of the genitals between women.
5. **Tribidist** (*trib ah dist*)—Intimate homosexual relationships between females; the active female who assumes the male character towards the female partner in the sexual act is called a tribidist.
6. **Urophilia**, homosexual—A sexual use of urine or urination or defecation. **Slang**: Shit and piss given.

CHAPTER 19

Fetishes

INTRODUCTION

The word "fetish," the Portuguese for "charm" is not often an attachment to things which are *not normally sexual objects,* and often to the exclusion of normal sexual relationships. The most common fetishistic objects are clothes, especially undergarments and shoes, furs, rubber garments, manikins, leather handbags, cigarettes, towels, etc. Most often the object of a fetish (such as all of the above) is associated in some way with people or at times, animals. Many fetishists marry but are potent only if the wife takes part in the fetish (called *sexual piquantism*). In the full sense of the word, this is where something replaces the original love object by the process of association conditioning. Also, fetish refers to situations in which one's early sexual feelings become fixated on some object. In a true fetish, sexual gratification is usually possible only if the fetish object (which can be anything from underwear to animals to body parts) is present, although sometimes a mental picture is satisfactory.

Early students of human sexual behavior observed that some individuals were capable of being aroused sexually only when they viewed or engaged in tactile contact with only part of their partner's body other than the genitalia. This is true even when they viewed or engaged in tactile contact with certain inanimate objects. Many men could not become aroused sexually except by fondling the breasts of a woman, by touching her feet or by stroking her undergarments or other such specific behavior. Other men are capable of sexual arousal and sexual satisfaction in normal ways but were nonetheless more quickly aroused and/or more easily satisfied when certain nongenital bodily parts or inanimate objects were viewed or touched.

To describe the condition in which an individual's sexual arousal and/or satisfaction was related in some way to viewing or touching a nongenital bodily part or an inanimate object, these early writers employed the term "fetishism," which is derived from the Latin, *factitious,* i.e., "made by art," ergo, "artificial"—the implication being that such arousal and/or satisfaction was "unnatural." This term definition proved problematic because an overwhelming majority of both males and females traditionally find themselves more easily aroused by certain bodily parts than others. Many males, for example, experience intense arousal as a result of viewing or touching female breasts, but are not especially excited by female buttocks, while other men have the response. Likewise, many females are aroused by viewing or touching male biceps, thighs, or pectoral muscles, while others are unexcited by this but are aroused by viewing or touching other body parts. Thus, fetishism, as defined by the early writers, could be seen as being an all but universal phenomenon. Thus today, concern exists only if it totally replaces normal sexual responses and/or if it harms the other person; or it is an object normally far removed from normal sexuality, such as a towel or excretion.

THE MAIN TYPES OF FETISHES

1. **Abasiophilia** (*Greek, abasio* = lameness) is a type of paraphilia, a specific sexual attraction to handicapped persons who must use orthopedic braces such as leg braces or even wheelchairs.
2. **Anaclitism** involves deriving sexual arousal from objects that one was exposed to as an infant such as diapers. Also called **diaper fetish**, **diaperism**, **autonepiophilia**. Slang is **adult babyism**. **analphiliac** or **analerotic** (*ay nal er* otic)—Any erotic attachment to or focus on the anus.
3. **Bestiality** (*beast te al a ty*) (*Middle English, beasts* = animal)—Specifically sexual intercourse with animals or a drive or desire to have sex with an animal.
4. **Coprolangia** (*kop ro lag e ah*) (*Greek, kopros* = dung)—One who becomes sexually excited from the taste and/or smell of sweat, and/or feces. If the fetish involves feces only, the person is called a **coprolangniac**.
5. **Coprolalia** (*kop rho lay le ah*) (*copro* = dirty; *lalia* = language)—Deriving sexual stimulation from using obscene, unacceptable language, especially that related to excretion or sex, writing obscene words or sentences on walls, or writing obscene letters. Also called a **word fetish** or **coprophrasia**. A person with this disorder is called a **coprolaliac**.
6. **Coprology** (*kop rol ah je*) (*Greek, kopros* = dung; *logy* = the study of)—The scientific/literary study of both hard- and soft-core pornography in both art and literature.

7. **Coprophilia** (*kop re fili e*) (*Greek, kopros* = dung; *philia* = love)—Sexual attraction to feces, the process of excreting or watching someone excrete. Both sight, sound and smell are often involved. A person with this fetish is a **coprophiliac**. **Slang**: pound cake. A person who goes further than a **coprophiliac**.
8. **Coprophagia**—One who actually consumes (eats) the excrement, the person is referred to as a **coprophagiac**.
9. **Depilation, sexual** (*dep ah le ah shion*)—The removal of pubic hair, usually by shaving, for sexual reasons, or a fetish for this behavior or appearance. Relates to and is a form of pedophilia. **Slang**: mowed lawn, clean box, round mound.
10. **Ecouteurism** (*e coun tour izm*)—One who obtains a high level of erotic satisfaction from listening to the sexual experiences or exploits of others—generally the more vivid and graphic, the better.
11. **Eonism**—A person who experiences sexual arousal from wearing the underwear of the opposite sex, often in private. This classification excludes transvestites, who enjoy dressing and make-up, as women dress, and public display of such.
12. **Ennumiphilia** (*in new me filia*)—A clothes fetish, often specifically for female underwear, bras, and such, although it could involve any clothing items, even shoes, shirts, pants, etc. Embarrassment related to buying them may cause the person to steal the items from someone's clothesline, or to shop lift from a store. Laundromats are now a common source of women's clothes.
13. **Erotolalia**—An abnormal drive and pleasure from uttering sexual words, terms, or phrases.
14. **Euodiaphilia** (*yoo oh dee' ah flil le ah*)—One who is erotically stimulated by smell, usually human body odors, especially those produced by the sexual organs, but also sweat, underarm, and feet odors. They may obtain the underwear of their lovers, or any female, and use its smell as a source of stimuli in order to masturbate. Also called: **osmolagnia** and **osphresiologniac**. A person who has this condition is a euodiaphiliac.
15. **Gluteophilia** (*glu te oh fil le ah*) (*Greek, gloutoi* = buttocks)—A fetish involving the human buttocks, a buttock fetish, where the main or primary sexual stimuli is the human buttocks, usually those of the opposite sex. A form of partialism. **Vernacular**: (*French derrier* = behind). **Slang**: ass man, butt freak, cheek man, hind end man, keester man, moon shot lover, posterior freak, rear endlover.
16. **Graphophilia** (*graf oh fil le ah*) (*Greek, grapho* = writing; *philia* = love)—An abnormal attachment to reading matter that vividly and graphically describes

sexual experiences; any love of pornographic writing. A person with this fetish is a **graphophiliac**.

17. **Hair fetish**—The love of hair; often involves the act of a person cutting off and saving the hair of other persons, usually of the opposite sex. A once common place to practice this deviation was indoor movie theaters. The person, usually a male, sits behind a young girl who has long hair hanging over the seat, then cuts a large lock of it, takes it home, and then masturbates, using both the hair and his thoughts of how it was obtained as stimuli.
18. **Haro Sexualis Partalis** (*har oh sex you alis part ah lis*)—A sexual drive for certain colors on a person, includes colors of hair, eyeglasses, etc.
19. **Hyphephilia** (*hi fa fil le ah*)— a high level of sexual pleasure from touching fabrics, cloth, and other pleasurable surfaces. This term includes everything from sensual touching to a fetish for touching. Males with this fetish will go out of their way to touch the clothes, especially undergarments of women, often without their permission or in occupations that allow them to do this, such as a clothes tailor.
20. **Infantilism** (*in fan til izm*)—Sexual gratification to dress and act like a baby. Often includes wearing diapers, etc., and sometimes even excreting in them.
21. **Kiromah fetish** (*ki ro mah*)—a sexual fetish that makes heavy use of towels or wash cloths. Often the person masturbates while using the towel, which is often one stolen from a woman that he personally knows or even from a stranger's laundry or from a laundromat.
22. **Kleptomania, sexual** (*klep to may ne ah*) (*Greek, kleptes* = steal; *mania* = compulsion)—The compulsive desire to steal sexually appropriate articles, such as women's underwear, often for sexual reasons, sometimes to the point of orgasm. Both the article, the act of stealing, and imaging doing so are involved. A person who engages in this behavior is grouped with all compulsive stealers, collectively called **kleptomaniacs**. Sometimes the stealing of any object may produce a sexual response, although most types of kleptomania typically involve stealing for neurotic, not sexual, reasons.
23. **Mannikinism** (*man na ken izm*)—A strong, persistent sexual desire for a mannequin (a plastic statue used to display clothes in a store); a mannequin fetish. A person who engages in this behavior is a **mannikinmaniac**.
24. **Mixoscopia** (*mik so sko pe ah*)—A compulsion to watch others perform sexually, sometimes especially if they are forced to perform. As the watching itself is what sexually excites the watcher, sometimes a couple is forced to perform. The person may even break in a house and force a husband and wife to perform intercourse in front of him. Another, less deviant form, is where a person forces his own lover or wife to perform sex with another person, the

mixoscopic (the person with this problem) often achieving orgasm from watching. This deviancy is considerably different from a voyeurist.

25. **Mysophilia** (*mi so fil leah*)—A fetish in which the person finds highly erotic specific behavior that is revolting to most people, such as eating vomit. Mild forms include smelling soiled underwear or obtaining used menstrual pads or even sweaty clothes to smell. Urophilia and coprophilia are examples, both related to corprophilia. A person with this condition is called a mysophiliac.
26. **Narcissophilia** (*nar ciss oh file ah*)—When one's major erotic orientation is mentally and physically only from oneself, often both as a person and physically. One may enjoy watching one's self nude, take nude pictures of oneself, or admire oneself in a mirror. A person with this problem is a **Narcissophiliac**. Sexual arousal and/or pleasure is experienced by also contemplating one's own attractiveness, wholly without altererotic physical or psychic stimulation, and regularly or exclusively aroused sexually by contemplating his/her own attractiveness, and wholly without altererotic physical or psychic stimulation. The term derives from the name of Narcissus, the character in Greek mythology who, upon seeing his reflection in a pool of water, was so overcome with emotion, that when he attempted to kiss the reflection, fell into the pool and drowned. Other form: Narcism, sexual.
27. **Necrofetish** (*nekro fet ish*)—One who enjoys looking at or touching the dead for sexual reasons; a fetish or compulsion for corpses, or a liking to be near the dead, especially to look at or touch the dead.
28. **Olfactory fetish**—See Euadiaphilia above.
29. **Partialism** (*par tile izm*)—The condition where one is obsessed or highly attracted to one specific part of the body to the exclusion of the other parts. During sex or when looking at a person of the opposite sex, the partialists devotes most of his/her energy and attention to this part of the body to the exclusion of all other parts. The attachment could be to the breasts, buttocks, nape of the neck, legs or another specific body part.
30. **Photophilia** (*foto fil le ah*) (*Greek, photo* = picture; *philia* = love)—A strong erotic attachment or attraction to one dimensional photographs or printed pictures; a male who enjoys looking at, usually pictures of nude women, largely to the exclusion of other types of sexual stimuli. Some photophiliacs collect pictures of nudes, accumulating multi-thousands of magazines, films, and other stimuli. A person with this orientation is a **photophiliac**.
31. **Pictophilia**—A strong need for, or a dependency on, erotic photographs for sexual stimuli. Similar to pornophilia except the focus is only on the visual stimuli of pictures.

32. **Pornolagnia** (*porno lag ne ah*)—A morbid, powerful sexual interest in prostitutes, especially behavior characterized by the inability to achieve sexual arousal and/or pleasure with any individual other than a prostitute.
33. **Pornophilia** (*por no fil iah ah*) (*Greek, porno* = sexual; *philia* = love)—A sexual deviation, usually in males, in which a person is erotically stimulated primarily by looking at or reading pornography to the exclusion of sexual involvement with living persons. Often these persons are not able to involve them in sexual relations with real people. Sometimes refers to a preoccupation with pornography, especially an abnormal attachment to pictorial pornography. If the person is attached primarily to written pornographic narratives, it is called **narratophilia**. A person with the picture fetish is a **pornophiliac**, one with a fetish for sexual stories is a **narratophiliac**.
34. **Pornographic fetish**—When one prefers pornography to living persons as a source of erotic stimuli. Often only pornography will sexually satisfy the person, and in order to have sex with a person, viewing pornography is often a necessary prelude or a required concurrent activity.
35. **Pygmalphilia** (*pig mal fel le ah*)—In Greek mythology, Pygmalion was a sculptor who fell in love with one of his own creations, a nude marble statue of a woman (he prayed to Aphrodite to animate it, his prayer was answered, and he married the girl). This condition refers to sexual desire or pleasure from a statue or statues, a statue fetish. Also called **icolagnia**, and the victim a **icolagniac**. A person with this is a **pygmalphiliac**. The term pygmalionism also refers to romantic an/or sexual attraction to an object in which the actor himself had a part in creating, especially romantic and/or sexual attraction on the part of a male for a younger female whom he has tutored and coached in the development of attractiveness, charm, or some specific skill. The title character in George Bernard Shaw's play, Pygmalion, who fell in love with a girl whom he had schooled in the social graces, is an example.
36. **Retifism** (*rat ah fizm*)—A foot fetish, or an abnormal attraction to feet, looking at, touching or rubbing them, etc.
37. **Saliromania** a obsessive sexual fetish involving soiling or disheveling sexual objects, usually an attractive person. Harming or injuring is not involved. Examples include tearing or damaging their clothing or more common covering then in mud or messing up their make-up. Mud wrestling is a common example. Also can be a form of sadomasochism.
38. **Salirophilia** a non-obsessive sexual fetish involving soiling or disheveling sexual objects, usually an attractive person. Not a form of sadomasochism.

39. **Scatophilia** (*skat oh fil le ah*)—Sexual feelings from touching or playing with excretion, either one's own or one's partners. **Slang**: A scat freak. It is also called **scatology**. A person with this condition is a **scatophiliac**.
40. **Somnophilia** (*Latin, somnus* = sleep and *Greek, philia* = love) The desire to sexually kiss or fondle someone (often a stranger) or attempting to have sex with one who is asleep. A person with this condition is a **somnophiliac**. **Slang**: sleeping princess syndrome.
41. **Stethophilia** (*steth oh fel le' ah*) (*Greek, stetho* = bosom or breast)—A breast fetish, or the strong sexual attraction primarily to the breasts, usually males that are erotically attracted to female breasts, especially large breasts (but the preference may vary from flat women to women with large pendulous breasts, but usually firm, round breasts where the nipple is in the center and points straight out from the body is preferred). A person with this fetish is a **stethophiliac**.
42. **Trichophila** one who becomes sexually aroused by human hair and may involve certain hairstyles, colors, textures, or sheen variations. It may involve looking, touching, or watching someone else touch a person's hair, usually a woman. Often very long hair is involved.
43. **Undinism** (*un din ism*)—Sexual arousal from the contemplation of contact with water, urine, or urination.
44. **Urolagnia** (*yor oh lag ne ah*) (*Greek, uro* = urine)—A condition where the person is sexually aroused by the **sight** and **odor** specifically of urine. This fetish is often found in sado-massochists. A person with this fetish is a **urolangniac**. The person may steal or buy underwear soaked with urine, or that which has a urine smell. Children not uncommonly play games related to urinating (who can shoot the farthest), and urinals are designed to reduce inhibitions of urinating together—women most always have more privacy.
45. **Urophilia** (*ur oh fhil e ah*) (*Greek, uro* = urine; *philia* = love)—urine lover, or where the person is sexually excited by urine (usually human) and/or the process of urination (see also coprolagnia and urolagnia). Often sight, smell, and sound are all involved. A person with this fetish is a **urophiliac**. One who enjoys being urinated or excreted on, is called a *Toilet Slave*. **Slang**: the process of urinating on someone else is called *Golden Showers* or *water sports*.
46. **Zoophilia** (*zoo fil e ah*) (*Greek, zoo* = animals; *philia* = love)—An erotic attachment, desire or sexual relationship with animals, usually of a specific type; a passion for animals, often of a fetish nature. It may involve genital, oral, or anal sexual behavior. The person may also masturbate the animal, or the animal may perform fellatio on the person with this fetish, called a **zoophiliac**. Zoophilia is both a sexual attraction to or any erotic contact with

animals, in contrast to bestiality, which is sexual intercourse only, sometimes without erotic attachment to the animal. Also called **zooerastia** (*zoo er as te ah*). A morbid, powerful zoophilia is called a **zooeraghia**.

47. **Zoophilism**—Sexual pleasure from nongenital contact with animals.
48. **Zoonophilia** (*zoon oh fil e ah*)—An erotic attachment to a non-human, non-animal creature, i.e., of one from another planet, solar system, or galaxy. Creatures who engage in sex with human females are an **androzoon**.

CHAPTER 20

Sadomasochism

Sadomasochism (*sad oh mas oh kizm*) is a person who is both a sadist and a masochist; the co-existence of a high degree of submissive and aggressive behavior in the same person, usually it is either person specific or alternates between persons. **Slang:** S & M; English Culture, discipline or "into discipline" or a "stem disciplinarian". Types of sadomasochism include the following.

1. **Algolognia** (*al go lag ne ah*) (pain lust)—**Active algolagnia** is an enjoyment or a pleasure from causing physical or psychological pain to others (see sadism) **Passive algolagnia** is a condition with masochistic elements or where the person also feels pleasure from physical pain (see masochism). A person who derives sexual pleasure from any type of pain, although it is usually physical as opposed to psychological , is called an **algologniac**.
2. **Sadism**—The need to cause pain, to physically punish and/or otherwise to humiliate one's sexual partner. Basic classifications of sadism include:
    a. **Active sadism**—The drive to act out one's own desire on others. It is the opposite of **ideal sadism** in which no effort is made to act on one's thoughts in this area.
    b. **Actrotomophilia** (*act tro to mo filia*)—Fantasizing or desiring erotic contact with an amputee, a person that has lost a leg, arm, or both. One with this drive is an **actrotomophiliac**.
    c. **Apotemnophilia** (*ah po tem mo filea*)—Fantasizing oneself as an amputee for sexual reasons. Some persons actually try to become an amputee for sexual reasons, and some succeed. The person with this compulsion is called a **apotemnophiac**.
    d. **Asphyxiophilia** (*as fix oh fel e' ah*)—A drive for the erotic responses that can be obtained from partly choking oneself, so as to reduce the body's supply of oxygen, and consequently increase sexual pleasure. A person

who engages in this behavior is an **asphyxiophiliac**. Also called **autoerotic asphyxia**. **Slang**: scarfing.

e. **Auto-sadism**—A type of sadism in which the subject enjoys inflicting pain on him or herself. An example is a case recorded by Wertham of a man who forced hundreds of straight pins in his body in between his legs. The person who practices this is an **auto-sadist**.

f. **Autoassinatophilia** (*auto ah sass in ah to fieah*)—An erotic response from staging or imagining one's own homicide. The person with this compulsion is an **autoasinatophiliac**.

g. **Depilation**, **sexual** (*de pill e ay shion*)—A form of sexual behavior involving the shaving of pubic hair, either from the act, the result, or looking at the result; it may have sadomasochism elements.

h. **Dippoldism** (*dip pol dizm*)—An adult who achieves sexual pleasure from the flagelation, beating, or cruel treatment of a child, often a child from four to ten years of age.

i. **Dacryphilia** where pleasure is derived from seeing someone of the opposite sex in strong, often negative, emotional states, often involving tears. A man may seek out women to terrorize for the satisfaction and may or may not rape them. In some cases, the end result can be murder.

j. **Ecstasy Intoxication** (*eks tah see in tox ah ka shon*)—It is the point reached when the pain level for a sadist has reached its zenith of effectiveness, or when the sensation of pain is suppressed by stronger urges for more pain or by a sensation of sexual desire that results from the ability to tolerate a great deal of pain. It is at this point that major destruction of body tissues occurs and, at times, serious injury.

k. **Homicidophilia** (*hom ah ci do fil e ah*)—The drive to murder or fantasize committing homicide, largely or totally for erotic reasons, or this drive is the person's primary sexual stimulus. A person with this condition is called a **homicidophiliac**. It is, fortunately, very rare. See also necro-sadism.

l. **Ideal Sadism** (*ide al sad izm*)—The case in which sadistic acts exist solely in the imagination of the subject. It may precede actual sadistic acts or may be a stage toward developing active sadism. It is directly opposite of **active sadism**. Persons with this problem are **ideal sadists**.

m. **Pederosis** (*ped er oh sis*)—A sadistic fetish for the posterior of young boys. The offender may "spank" or beat or simply pinch young children in order to achieve orgasm or sexual feelings. It is also called **Podes Complex**.

n. **Piquer** (*pe koor*) (*French, piquer* = to prick or sting)—A type of sadist who stabs or "sticks" their victims, usually girls or women, with ice picks or other sharp instruments. Usually males, and often in a crowd, the piquer will stab a woman in the buttocks or breast, then immediately move away so as not to get caught, but will stay close enough so as to watch the victim in order to enjoy her expression or "show" of pain, surprise, or other reactions, this often being the primary source of gratification from the act.

o. **Sadism** (*sad izm*)—After Marquis De Sade (1740–1814) whose writings describe various painful sexual aberrations. The term thus describes those who obtain sexual pleasure from physically hurting or torturing others. It may involve ill-treatment, verbal humiliation, and/or misuse of the love object (person) or physical torture, especially inflicting pain on the genital organs or breasts. A person who involves him or herself in this behavior is a **sadist**. **Slang**: G/T.

p. **Sadistic pedophilia**—A type of pedophilia in which an adult inflicts pain or tortures a child. The person with this condition is a **sadistic pedophiliac**.

q. **Symbolic Sadism** (*sim boll ik sad izm*)—The condition in which the subject expresses his sadistic impulses and gains sexual pleasure by acts of mutilation of those symbols that have specific meaning for him. This could include behavior that ranges from throwing darts at nude pictures, to writing certain words (names of past lovers) on some object and then defacing it. A person who does this is a **symbol sadist**. This is a common behavior in western society, is rarely a problem, and is considered a deviant act only in extreme forms.

r. **Tightlacing** involves wearing a tight corset to achieve extreme modifications of the waste, usually practiced by women, as can achieve as small as a 40 cm. waist. It usually distorts the ribcage and internal organs.

s. **Vampirism** (*vam pi er izm*)—A form of sadism where the person draws blood from a sex object, usually by biting, but sometimes by cutting, but always primarily for sexual reasons. A person who commits vampirism is a **vampirist**.

t. **Zoosadist**—One who derives sexual pleasure or arousal from inflicting pain upon animals. The behavior is called **zoo sadism**. **Slang**: animal hit or hitter.

3. **Masochism** (*mass oh kizm*)—is the desire and enjoyment from experiencing pain and suffering for sexual reasons. The masochist may be a homosexual or a heterosexual. A person involved in this is a **masochist**. Types include the following:

a. **Bondage and discipline**—Sexual pleasure from being tied up and being actively or passively flagellated. Also includes sex play involving being tied up, both painful and painless types. The psychological elements are important. **Slang**: B & D., to flog, leather games, leather. The following fetish elements are often involved
   i. **Leatherites**—Persons who prefer leather ties, boots, etc., in their bondage activities.
   ii. **Rubberites**—Persons who prefer rubber ties, straps, etc., in their bondage activities.

b. **Flagellation** (*fla jel lay shun*)—A passion for whipping, deriving sexual passion by whipping or being whipped, often to the point of bleeding. Either the person is not capable of orgasm without this element or this element is an important part of sex. Flagellation is usually performed by a woman, most often by whipping the male trunk, buttocks, or breasts. Self-flagellation is a cultural practice in some societies and is practiced in some places in groups as a religious or tribal rite. According to psychoanalysis, the whip is the symbol of the penis, and in the recipient the contortions induced from the pain of being whipped resemble an orgasm. Much erotic literature deals with flagellation, and it is said to be a sexual orientation more common among the English than other nations.

CHAPTER 21

Sexual Behaviors Involving Elements of Sadomasochism

The main type is **frotteurism** (*fro ter ism*) (*French*, one who rubs)—A sexual offense in which the offender, usually a male, presses his genitals against another person, usually a woman, but sometimes a male, and often in a crowd. He usually presses against the woman's buttocks and sometimes places his hands in his pockets to manipulate his genitals. He may even cut or remove the pockets to better manipulate his sex organs. Commonly involved in this condition are homosexual elements. The term also applies to any sexual excitement from rubbing one's genitals against another person, behavior called **frottage**. The other person is usually unwilling and often not aware of the offender's sexual excitement. A person who does this is a **frotteur**. **Slang**: a masher.

TYPES OF FROTTERISM

1. **Ear pincher**—When one follows women around and pinches their ears.
2. **Leg kisser**—A sexual perversion where a male typically crawls around under the seats in movie houses or other dark public places and kisses women's legs.
3. **Necrophagia** (*nec rho fay ge ah*)—The eating of other humans or ingestion of human flesh for sexual reasons, often children; cannibalism for sexual reasons. Another term is **anthropophagy** (*an thro po fay ge*) (human eater). Well known cases include Periandrus, Herod, Charlemagne, and Albert Fish. A person who commits necrophagia is a **nerophagiac**.
4. **Necrophilia** (*nek ro fil ea ah*; the British spelling is nekrophilia)—Literally a corpse lover; an abnormal attraction, especially an erotic attraction, to the dead or corpses; one who desires sexual relations (coitus) with the dead. They often

enjoy dressing, fixing the hair, etc. of the corpses, as well as undressing, caressing, and having oral, anal, or vaginal intercourse with it. To reduce the likelihood of the embalmers or others having sex with corpses laws in some countries require the corpse of attractive young girls to be first stored for several days to a week before anything is done with it. A person with powerful necrophilia drives is a **necromaniac**. See also **homicidophilia**. A person who commits necrophilia is a **necrophiliac**.

5. **Necro-sadism** (*nek ro sad* izm)—Lust murder, one who commits homicide and not uncommonly then mutilates the corpse for sexual reasons. A person who commits necro-sadism is a **necro-sadist**. They may mostly desire, and sometimes perform, primarily wonton mutilations of the corpse.
6. **Phyrophilia** (*Greek, pyro* = fire; *philia* = love—One who receives erotic enjoyment from setting fires and then watching them burn. They typically enjoy primarily the excitement of the crowd, persons screaming, and the general excitement that is most always attendant with fires. The basic types of phyrophilia include:
    a. **Sexual incendiarism** (*sex you al in sen di a ri izm*) (*Latin, incendiary* = setting on fire)—Willful destruction of property by fire for reasons that involve sexual elements.
    b. **Fire-water complex**—a common element of psychopathic sexual incendiarists. After lighting a fire (pyromania) a period of exhibition occurs, followed by a desire to urinate or excrete.
7. **Rape** (*Latin, rapeve* = to take by force)—Having coitus by force, without the victim's full permission. Often the victim is related or known to the offender. Also, **statutory coitus** or sexual intercourse by an adult with a person under the age of consent, usually 16 years old. A person who rapes is a rapist. Usually, sadomasochistic elements are involved.
8. **Biastophilia** (*Greek, biastes* = rape) where sexual arousal is dependent on sexually assaulting a non-consenting person, often a stranger. Common in serial killers and may not involve sexual intercourse.
9. **Rapophilia** (*rape oh fil e ah*)—A need, in order to achieve orgasm, or great erotic joy for the obviously terrified resistance of non-consenting strangers.
10. **Sabotour** (*sab ah tour*)—Sexual offense where the offender cuts the trousers off his victim.
11. **Saliromania** (*Greek, salir* = soiling; *mania* = mad craving)—Achieving erotic excitement by deliberately destroying or soiling an object, usually urinating on a woman's clothes. Also includes behavior such as damaging or besmirching nude statues or paintings. A person who does this is a **saliromaniac**.
12. **Vorarephilia** where arousal occurs from the idea of being eaten.

13. **Xenophily** (*Greek, xeno* = foreign or unknown)—Arousal from strangers, often a factor in rape, but most **xenophiliacs** are not rapists. Often, they will pick up women in bars, have sex with them, and never want to see them again.

CHAPTER 22

Erogenous Zones

Erogenous Zones are **Sexual Body Parts**, the areas of the body that, from tactile stimulation, lead to sexual arousal. Also called privates or hit areas. All body skin responds to some degree, but some are far more responsive than others.

1. **The female breasts**—-mammary glands—-non-development in adult females due to hormone or genetic causes is called amastia. **Slang words** for female breasts: balloons, bazooms, bee stings, big grownies, boobies, boobs, brace and bets, bristols (British), bubbies, bubs, buds, buffers, bust, chubbies, broopers, fore-buttocks, globes, grapefruits, headlights, jugs, kazongas, knockers, knobs, lungs, mammae, manchesters, maracas, melons, milkers, milk wagons, mountains, muffs, muffins, nay-nays, nubbies, pair, silicones, tabs, teats, tits (or titties), topsides, twin hills, twin peaks, upper deck, voos.
2. **The male sexual organ**—penis—**Slang words**: almond rock, banana, bar, bean, bent stick, big brother, blind Jack, blind Bob, bone bowsprit, box opener, butcher knife, canasta, candy bar, Charlie, cheese cutter, chopper, cock, cookie, dark meat, dead meat, dick, dingus, dink, dirty barrel, do jigger, doodad, dong, dork, fancy work, fuck stick, gadget, giggle stick, giggling pin, golden rivet, golden rod, goober, green thumb, groceries, gun, hammer hidden treasure, hand-made, hand-reared, he, honey stick, hot dog, impudence, Jack-in -the-box, jang, jerking iron, jing-jang, jock, Johny, Johnson, joint, joy knob, joy stick, knob, linga, lingam, little brother, lollipop, long Hohn, lower deck, matrimonial peacemaker, meat, meat whistle, meat with two vegetables, membrum virile (Latin), Mickey, middle leg, mortar and pestle, muscle, mutton dagger, old blind, old thing, one-eyed monster, organ, pax wax, pecker, pencil, Peter, phallus, piccolo, pintle, piston rod, pogo stick, poker, pole, poontanger, pork chopper, pork sword, priapus, prick, privates, prong, pudenda, pulse, putz, red cap, red-hot poker, rhubarb, Richard, rod Roger,

Rupert, rusty rifle, schmuck, schnitzel, schong, schwanz, sexing piece, six inches, skin flute, snake, spout, stalk, stick, sticker, stinger, stuff, swagger-stick, swanska, sword, tadger, tallywhacker, thing, Tommy, tonge, tongue, tool, wang, weapon, weenie, wee-wee, whang, whelp, whistle, white meat, wick, wiener, Willie-whacker, wrinkle, wire, worm, yang, yard, ying-yang, yo-yo, zubrick. A large penis is a **horsechoker**. The penis and testicles are the **virilia**. **Slang** words for the testicles: balls, ballocks (old term).

3. **Vagina**—The female sexual organs. The external parts are collectively called the **vulva**. Synonyms for the vagina: bearded clam, bearded lady, black velvet, box, bread, bun bush, bushy park, business part, cabbage, cake, canasta, chuff, cockpit, conch, cooch, cookie, crack, cunnus, cunt, cush, cut, cuzzy, dark meat, dead end street, dicky, dido, dirty barrel, dog's mouth or nose, fort bushy, fur, furburger, fur pie, futy, garden, gash, geography, gig, gigi, gonad, groceries, growl, hair pie, hairy ring, hairy wheel, happy valley, hatchi, hidden treasure, hole, honey pot, hot box, Jack-in-a-box, jaxy, jelly, jelly roll, jung-jang, joxy, little sister, lollipip, lower deck, meat mortar and pestle, moustache, mowed lawn, muff, nookie, old thing, one that bites, organ, piece of ass, pink poontang, pudendum, puka, punce, pussy, quiff, quim, rattlesnake canyon, rubyfruit jungle, scratch, she, slash, slit, slot, snake pit, snatch, snippey, split, tail, thing, toolbox, twat, twot, white meat, wick burner, yoni, zosh.
4. Other body parts
    a. **Clitoris**—The source of female organism. **Slang**: clit, twig.
    b. **Public Hair**—In a female termed bush, beaver, beard, mound of Venus.
    c. **Anus**—The posterior opening of the large intestine. **Slang**: little rose, the flower.
    d. **Buttocks**—**Slang**: cheeks.
    e. **Creme de la Creme**—The sexual lubricating fluid, produced by the vagina.
    f. **Umbilicus**—The navel.

CHAPTER 23

Sexual Phobias or Repulsions

Many phobias involving sex exist. A phobia is an irrational fear of some object or event caused by a bad experience with the object or event causing pain or fear. The most common sexual phobia by far is homophobia, fear of homosexual sex or, more commonly, a fear of being around homosexuals due to being sexually molested as a young boy by a male. The fear often generalizes into a fear of males in general. The vast majority involve males, but occasionally a female may molest a young girl or occasionally a young boy.

A 2004 research study by the John Jay College of Criminal Justice for the United States Conference of Catholic Bishops, identified 4,392 Catholic priests and deacons in active ministry between 1950 and 2002 that have been accused of under-age sexual abuse of 10,667 individuals. The church openly accepted homosexuals in the ministry because all priests, both gay and straight, took vows to be celibate. Thus, their sexuality was not a concern. The cost to the church so far was over four billion dollars. The Boy Scouts have faced the same problem. A total of 7,819 male volunteers were removed due to evidence that they sexually abused male children, forcing the Boy Scouts into bankruptcy.

COMMON EXAMPLES OF SEXUAL PHOBIAS

1. **Androphobia** (*an dro fo bee ah*)—An intense dislike or fear of men by a female, and one which usually precludes normal sexual relations with males.
2. **Erotophobia**—A total dislike of anything erotic, sex and related.
3. **Gynephobia**—Morbid fear or dislike of women.
4. **Haphenphobia**—Morbid fear of being touched.
5. **Misandria**—A morbid fear of men by women.
6. **Misogamy**—Fear of marriage.

7. **Misogynist**—Fear of women by a man.

Bibliography

Abraham, Jerrold L. 1980. "Medical aspects of homosexuality." *New England Journal of Medicine* **302**(8):463-464, February 21.

Adams, Bob. 2001. "Typhoid Larry." *The Advocate*, June 19.

Ahmad, S., and A. Sukthankar. 1998. "Gonorrhea and chlamydia coinfection in gay men." *International Journal of STD and AIDS* **9**(1):59-61, January.

Averbach, John. 2000. "Gay and lesbian health in the big city." *The Advocate*, June 20, p. 20.

Bagemihl, Bruce. 1999. *Biological Exuberance: Animal Homosexuality and Natural Diversity*. New York, NY: St. Martin's Press.

Bell, Alan P., Martin S. Weinberg, and Sue K. Hammersmith. 1981. *Sexual Preference: Its Development in Men and Women*. Bloomington, IN: Indiana University Press.

Beral, Valeria, Diana Bull, Sarah Darby, Ian Weller, Chris Carne, Mick Beecham, and Harold Jaffe. 1992. "Risk of Kaposi's sarcoma and sexual practices associated with fecal contact in homosexual or bisexual men with AIDS." *The Lancet* **339**(8794):632-635, March 14.

Bernstein, K.T., R. Tulloch, J. Montes, G. Bolan, I.E. Dyer, M. Lawrence, A.P. Kaur, D. Kodagoda, H. Rotblatt, P. Kerndt, R. Gunn, P. Weismuller, and EIS officers, CDC. 2001. "Outbreak of syphilis among men who have sex with men—Southern California, 2000." *The Journal of the American Medical Association (JAMA)* **285**(10):1285-1287, February 23.

Biggar, Robert J., Mads Melbye, Peter Ebbesen, Dean L. Mann, James J. Goedert, Robert Weinstock, Douglas M. Strong, and William A. Blattner. 1984. "Low T-lymphocyte ratios in homosexual men: Epidemiologic evidence for a transmissible agent." *Journal of the American Medical Association* **251**(11):1441-1449, March 16.

Bolton, R. 1992. "AIDS and Promiscuity: Muddles in the Models of HIV Prevention." *Medical Anthropology* **14**(2-4):145-223, May.

Bounds, W. 1997. "Female condoms." *European Journal of Contraception and Reproductive Health Care* **2**(2):113-116, June.

Burke, L.K., and D.R. Follingstad. 1999. "Violence in lesbian and gay relationships: Theory, prevalence, and correlational factors." *Clinical Psychological Review* **19**(5):487-512, August.

Byne, William. 1994. "The biological evidence challenged." *Scientific American* **270**(5):50-54, May.

Cameron, Paul. 2002. "Homosexual partnerships and homosexual longevity: A replication." *Psychological Reports* **91**(2):671-678, October.

Cameron, P., K. Cameron, and K. Proctor. 1989. "Effect of homosexuality upon public health and social order." *Psychological Reports* **64**(3 Pt. 2):1167-1179, June.

Cameron, Paul, Kirk Cameron, and William Playfair. 1998. "Does homosexual activity shorten life?" *Psychological Reports* **83**(3 Pt. 1):847-866, December.

Cameron, Paul, William Playfair, and Stephen Wellum. 1994. "The longevity of homosexuals: Before and after the aids epidemic." *Omega* **29**(3):249-272, November.

Carey, R.F., W.A. Herman, S.M. Retta, J.E. Rinaldi, B.A. Herman, and T.W. 1992. "Effectiveness of latex condoms as a barrier to human immunodeficiency virus-sized particles under conditions of simulated use." *Sexually Transmitted Diseases* **19**(4):230-304, July-August.

Cheong, I., A. Lim, C. Lee, Z. Ibrahim, and K. Sarvanathan. 1997. "Epidemiology and clinical characteristics of HIV-infected patients in Kuala Lumpur." *Medical Journal of Malaysia* 52(4):313-317, December.

Chin, James. 2000. *Control of Communicable Diseases Manual.* Washington, D.C.: American Public Health Association.

Chrestiansen, M.A., and G.B. Lowhagen. 2000. "Sexually transmitted diseases and sexual behavior in men attending an outpatients' clinic for gay men in Gothenburg, Sweden." *Acta Dermato-Venereologica* **80**(2):136-139, March.

Christenson, B., C. Broström, M. Böttiger, J. Hermanson, O. Weiland, G. Ryd, J.V.R. Berg, and R. Sjöblom. 1982. "An epidemic outbreak of hepatitis A among homosexual men in Stockholm." *American Journal of Epidemiology* **116**(4):599-607, October.

Chu, Susan Y., Thomas A. Peterman, Lynda S. Doll, James W. Buehler, and James W. Curran. 1992. "AIDS in bisexual men in the United States: Epidemiology and transmission to women." *American Journal of Public Health* **82**(2):220-224, February.

Clark, William R. 1995. *At War Within: The Double-Edged Sword of Immunity.* New York, NY: Oxford University Press.

Colson, Charles, and Nancy Pearcey. 1999. *How Now Shall We Live?* Wheaton, IL: Tyndale.

Communicable Disease Report. 2000. "Increased transmission of syphilis in Brighton and Greater Manchester among men who have sex with men." *CDR Weekly* **10**(43):383, 386.

Cock, K.M., and H.A. Weiss. 2000. "The global epidemiology of HIV/AIDS." *Tropical Medicine and International Health* **5**(7):A3-9, July.

Cone, Dennis. 1994. "Homosexual promiscuity." *Current Thoughts and Trends* **10**(9):28.

Cooper, Al. 2000. "Cyber-Compulsive." *The Advocate* Magazine, June 20.

Corey, Lawrence, and King K. Holmes. 1980. "Sexual transmission of hepatitis A in homosexual men: Incidence and mechanism." *The New England Journal of Medicine* **302**(8):435-438, February 21.

Darrow, W.W. 1989. "Condom use and use-effectiveness in high-risk populations." *Sexually Transmitted Diseases* **16**(3):157-160.

Dauden, E., R. Martin, C. Feal, E. Munoz, and J. Fraga. 2000. "Eccrine ductal mucinosis in a human immunodeficiency virus-positive patient with probable scabies." *British Journal of Dermatology* **143**(6):1335-1336, December.

Davidovich, U., J. de Wit, N. Albrecht, R. Geskus, W. Stroebe, and R. Coutinho R. 2001. "Increase in the share of steady partners as a source of HIV infection: A 17-year study of seroconversion among gay men." *AIDS* **15**(10):1303-1308, July 6.

Davidovich, U., J.B. de Wit, and W. Stroebe. 2000. "Assessing sexual risk behavior of young gay men in primary relationships: The incorporation of negotiated safety and negotiated safety compliance." *AIDS* **14**(6):701-706, April 14.

Day, Lorraine. 1991. *AIDS: What the Government Isn't Telling You.* Palm Desert, CA: Rockford Press.

Deparis, X., R. Migliani, and M. Merlin. 1999. "Evidence of risk factors for condom breakage among French military personnel stationed overseas." *Medicine Trop/Cale* **59**(3):266-270, January.

Donovan, B. 1995. "Barriers to conception and disease." *Annals of the Academy of Medicine, Singapore* **24**(4):608-614, July.

Donovan, B. 1995a. "Condoms and the prevention of sexually transmissible diseases." *British Journal of Hospital Medicine* **54**(11):575-578, December.

Dooley, Samuel W., Margarita E. Villarino, Mercedes Lawrence, Louis Salinas, Samuel Amil, John V. Rullan, William R. Jarvis, Alan B. Bloch, and George M. Cauthen. 1992. "Nosocomial transmission of tuberculosis in a hospital unit for HIV-infected patients." *JAMA* **267**(19):2632-2634, May 20.

d'Oro, L.C., F. Parazzini, L. Naldi, and C. La Vecchia. 1994. "Barrier methods of contraception, spermicides, and sexually transmitted diseases: A review." *Genitourinary Medicine* **70**(6):410-417, December.

Duyves, M. 1993. "The minitel: The glittering future of a new invention." *Journal of Homosexuality* **25**(1-2):193-203.

Elford, Lee, G. Bolding, M. Maguire, and L. Sherr. 1999. "Sexual risk behavior among gay men in a relationship." *AIDS* **13**(11):1407-1411, July 30.

Ellis, Lee, and M.A. Ames. 1987. "Neurohormonal functioning and sexual orientation: A theory of homosexuality-heterosexuality." *Psychological Bulletin* **10**(2)233-258, March.

Faundes, A., C. Elias, C. Coggins. 1994. "Spermicides and barrier contraception." *Current Opinion in Obstetrics and Gynecology* **6**(6):552-558, December.

Feldblum, P.J., C.S. Morrison, R.E. Roddy, and W. Cates, Jr. 1995. "The effectiveness of barrier methods of contraception in preventing the spread of HIV." *AIDS* **9**(Suppl A):S85-93.

Feldblum, P.J., J.J. Bwayo, M. Kuyoh, M. Welsh, K.A. Ryan, and M. Chen-Mok. 2000. "The female condom and STDs: Design of a community intervention trial." *Annals of Epidemiology* **10**(6):339-346, August.

Feldblum, P.J., M.A. Kuyoh, J.J. Bwayo, M. Omari, E.L. Wong, K.G. Tweedy, and M.J. Welsh. 2001. "Female condom introduction and sexually transmitted infection prevalence: Results of a community intervention trial in Kenya." *AIDS* **15**(8):1037-1044, May 25.

Fox, Earle. 1994. "The Diseases of Homosexuality." *Emmaus News* **1**(37):2.

Fox, K.K., C. del Rio, K.K. Holmes, E.W. Hook III, J.S. Knapp, G.W. Procop, S.A. Wang, W.L. Whittington, and W.C. Levine. 2001. "Gonorrhea in the HIV era: A reversal in trends among men who have sex with men." *American Journal of Public Health* **91**(6):959-964, June.

Frisch, Morten, Bengt Glimelius, Adriaan J.C. Van Den Brule, Jan Wohlfahrt, Chris J.L.M. Meijer, Jan M.M. Walboomers, Sven Goldman, Christer Svensson, Hans-Olov Adami, and Mads Melbye. 1997. "Sexually transmitted infection as a cause of anal cancer." *The New England Journal of Medicine* **337**(19):1350-1358, November 6.

Goldenberg, Herbert, and Irene Goldberg. 1990. *Counseling Today's Families.* Pacific Grove, CA: Brooks/Cole.

Goldstone, Stephen. 2001. Editor of *Gayhealth.com* website.

Golombok, S., R. Harding, and J. Sheldon. 2001. "An evaluation of a thicker versus a standard condom with gay men." *AIDS* **15**(2):245-250, January 26.

Greenblatt, Robert B. 1963. *Search the Scriptures: A Physician Examines Medicine in the Bible.* Philadelphia, PA: J.B. Lippincott.

Grossman, Lisa. 2015. "Time to lay 'born this way' to rest." *New Scientist* **227**(3031):18-19, July 25.

Hegyi, E., V. Hegvi, and T. Danilla. 1997. "Long-term trends of the incidence of syphilis and gonorrhea in the Slovak Republic during 1947-1994, with special regard to young age groups." *Bratislavske Lekarske Listy* **98**(10):563-571, October 1.

Herrell, R., J. Goldberg, W.R. True, V. Ramakrishnan, M. Lyons, S. Eisen, and M.T. Tsuang. 1999. "Sexual orientation and suicidality: A co-twin control study in adult men." *Archives of General Psychiatry* **56**(10):867-874, October.

Horgan, John. 1995. "Gay genes, revisited: Doubts arise over research on the biology of homosexuality." *Scientific American*, November.

Interlinear Translation of the Greek Scriptures. 1985. Brooklyn, NY: International Bible Students Association.

Kafka, M.P., and J. Hennen. 1999. "The paraphilia-related disorders: An empirical investigation of nonparaphilic hypersexuality disorders in outpatient males." *Journal of Sex and Marital Therapy* **25**(4):305-319, October-December.

Koblin, Beryl A., Nancy A. Hessol, Ann G. Zauber, Patricia E. Taylor, Susan B. Buchbinder, Mitchell H. Katz, and Cladd E. Stevens. 1996. "Increased incidence of cancer among homosexual men, New York City and San Francisco, 1978-1990." *American Journal of Epidemiology* **144**(10):916-923, November 15.

Lemp, George F., Melissa Jones, Timothy A. Kellogg, Giuliano N. Nieri, Laura Anderson, David Withum, and Mitchell Katz. 1995. "HIV seroprevalence and risk behaviors among lesbians and bisexual women in San Francisco and Berkeley, California." *American Journal of Public Health* **85**(11):1549-1552, November.

LeVay, Simon, and Dean Hamer. 1994. "Evidence for biological influence in male homosexual behavior." *Scientific American* **270**(5):43-57, May.

Lorber, Bennett. 1996. "Are all diseases infectious?" *Annals of Internal Medicine* **125**(10):844-851, November 15.

Macaluso, M., J. Kelaghan, L. Artz, H. Austin, M. Fleenor, E.W. Hook, III, and T. Valappil. 1999. "Mechanical failure of the latex condom in a cohort of women at high STD risk." *Sexually Transmitted Diseases* **26**(8):450-458, September.

Mavligit, Giora M., Moshe Talpaz, Flora T. Hsia, Wendy Wong, Benjamin Lichtiger, Peter W. Mansell, and David M. Mumford. 1984. "Chronic immune stimulation by sperm alloantigens. Support for the hypothesis that spermatozoa induce immune dysregulation in homosexual males." *JAMA* **251**(2):237-241, January 13.

McCutcheon, Marc. 1989. *The Compass in Your Nose: And Other Astonishing Facts About Humans*. Los Angeles, CA: Jeremy P. Tarcher, Inc.

McKusick, Leon, William Horstman, and Thomas J. Coates. 1985. "AIDS and sexual behavior reported by gay men in San Francisco." *American Journal of Public Health* **75**(5):493-496, May.

McMillan, A., H. Young, and A. Moyes. 2000. "Rectal gonorrhea in homosexual men: Source of infection." *International Journal of STDS and AIDS* **11**(5):284-287, May.

McMillen, S.I., and Donald E. Stern. 2000. *None of These Diseases: The Bible's Health Secrets for the 21st Century*. Grand Rapids, MI: Fleming H. Revell.

McNeill, John J. 1976. *The Church and the Homosexual.* New York, NY: Simon and Schuster.

Meeker, Meg. 2002. *Epidemic: How Teen Sex is Killing Our Kids*. Washington, D.C.: LifeLine Press.

Mekonen, E., and A. Mekonen. 1999. "Breakage and slippage of condoms among users in north Gondar Province, Ethiopia." *East African Medical Journal* **76**(9):481-483, September.

Melbye, Mads, and Robert J. Biggar. 1992. "Interactions between persons at risk for AIDS and the general population in Denmark." *American Journal of Epidemiology* **135**(6):593-602, March 15.

Munoz-Perez, M.A., A. Rodriguez-Pichardo, and Martinez F. Camacho. 1998. "Sexually transmitted diseases in 1161 HIV-positive patients: A 38-month prospective study in Southern Spain." *Journal of the European Academy of Dermatology and Venereology* **11**(3):221-226.

Palefsky, J.M., E.A. Holly, M.L. Ralston, and N. Jay. 1998. "Prevalence and risk factors for human papillomavirus infection of the anal canal in human immunodeficiency virus (HIV)-positive and HIV-negative homosexual men." *Journal of Infectious Diseases* **177**(2):361-367, February.

Pauk, John, Meei-Li Huang, Scott J. Brodie, Anna Wald, David M. Koelle, Timothy Schacker, Connie Cellum, Stacy Selke, and Lawrence Corey. 2000. "Mucosal shedding of human herpesvirus 8 in men." *New England Journal of Medicine* **343**(19):1369-1377, November 9.

Penn, Robert. 1997. *The Gay Men's Wellness Guide.* New York, NY: Henry Holt.

Remis, R.S., A. Dufour, M. Alary, J. Vincellette, J. Otis, B. Masse, B. Turmel, R. LeClerc, R. Parent, and R. Lovoie. 2000. "Association of hepatitis B virus infection with other sexually transmitted infections in homosexual men." Omega Study Group. *American Journal of Public Health* **90**(10):1570-1574, October.

Rohde, Paul, John Noell, Linda Ochs, and John R. Seeley. 2001. "Depression, suicidal ideation, and STD-related risk in homeless older adolescents." *Journal of Adolescence* **24**:447-460, August.

Rosen, Emanuel H. 1998. "How to think like a shrink." *Psychology Today* **31**(5):54-59, September.

Rueda, Enrique. 1982. *The Homosexual Network: Private Lives and Public Policy.* Old Greenwich, CT: Devin Adair.

Ryan, D.P., C.C. Compton, and R.J. Mayer. 2000. "Carcinoma of the anal canal." *New England Journal of Medicine* **342**(11):792-798, March 16.

Santinover, Jeffrey. 1996. *Homosexuality and the Politics of Truth.* Grand Rapids, MI: Baker Publishing Group.

Scutchfield, F. Douglas, and William Kech. 1997. *Principles of Public Health Practice*. Albany, NY: Delmar Publishing.

Shalit, Peter. 1998. *Living Well: The Gay Man's Essential Health Guide*. Los Angeles, CA: Alyson Books.

Siker, Jeffrey.. 1994. "How to decide? Homosexual Christians, the Bible, and gentile inclusion." *Theology Today* **51**(2):219-234.

Singh, S., R. Prasad, and A. Mohanty. 1999. "High prevalence of sexually transmitted and blood-borne infections amongst the inmates of a district jail in Northern India." *International Journal of STD and AIDS* **10**(7):475-478, July.

Soards, Marion L. 1995. *Scripture and Homosexuality: Biblical Authority and the Church Today.* Louisville, KY: Westminster Press.

Sparrow, M.J. 1999. "Condom failure in women presenting for abortion." *New Zealand Medical Journal* **112**(1094):319-321, August 27.

Spruyt, A., M.J. Steiner, C. Joanis, L.H. Glover, C. Piedrahita, G. Alvarado, R. Ramos, C. Maglaya, and M. Cordero. 1998. "Identifying condom users at risk for breakage and slippage: Findings from three international sites." *American Journal of Public Health* **88**(2):239-244, February.

Stephenson, Joan. 2000. "HIV risk from oral sex higher than many realize." *JAMA* **283**(10):1279, March 8.

Stone, E., P. Heagerty, E. Vittinghoff, J.M. Douglas, Jr., B.A. Koblin, K.H. Mayer, C.L. Celum, M. Gross, G.E. Woody, M. Marmor, G.R. Seage, III, and S.P. Buchbinder. 1999. "Correlates of condom failure in a sexually active cohort of men who have sex with men." *Journal of Acquired Immune Deficiency Syndromes and Human* **20**(5):495-501, April 15.

Taylor, D.J., and R.C. Dominik. 1999. "Noninferiority testing in crossover trials with correlated binary outcomes and small event proportions with applications to the analysis of condom failure data." *Journal of Biopharmaceutical Statistics* **9**(2):365-377, May.

Thomsen, Russel J. 1974. *The Bible Book of Medical Wisdom.* Old Tappan, NJ: Fleming H. Revell.

Tyler, Donald E. 1994. *The Other Guy's Sperm: The Cause of Cancers and Other Diseases.* Ontario, OR: Discovery Books.

Van de Ven, P., D. Campbell, S. Kippax, G. Prestage, J. Crawford, D. Baxter, and D. Cooper. 1997. "Factors associated with unprotected anal intercourse in gay men's casual partnerships in Sydney, Australia." *AIDS Care* **9**(6):637-649, December.

van Heeringen, C., and J. Vincke. 2000. "Suicidal acts and ideation in homosexual and bisexual young people: A study of prevalence and risk factors *Social Psychiatry and Psychiatric Epidemiology* **35**(11):494-499, November.

Wabinga, H.R., D.M. Parkin, F. Wabwire-Mangen, and S. Nambooze. 2000. "Trends in cancer incidence in Kyadondo County, Uganda, 1960-1997." *British Journal of Cancer* **82**(9):1585-1592, April 4.

Whyte, B. 2000. "UNAIDS estimates that half the teenagers in some African countries will die of AIDS." *Bulletin of the World Health Organization* **78**(7):946, January.

Wolitski, R.J., Ronald O. Valdiserri, Paul H. Denning, and William C. Levine. 2001. "Are we headed for a resurgence of the HIV epidemic among men who have sex with men?" *American Journal of Public Health* **91**(6):883-888, June.

Wong, M.L., R.K. Chan, D. Koh, and S. Wee. 2000. "A prospective study on condom slippage and breakage among female brothel-based sex workers in Singapore." *Sexually Transmitted Diseases* **27**(4):208-214, April.

www.ingramcontent.com/pod-product-compliance
Ingram Content Group UK Ltd.
Pitfield, Milton Keynes, MK11 3LW, UK
UKHW062258290726
14090UKWH00017B/763